Integrated Water Resources Management, institutions and livelihoods under stress:
bottom-up perspectives from Zimbabwe

Integrated Water Resources Management, institutions and livelihoods under stress:
bottom-up perspectives from Zimbabwe

DISSERTATION

Submitted in fulfilment of the requirements of

the Board for Doctorates of Delft University of Technology

and of the Academic Board of the UNESCO-IHE

Institute for Water Education

for the Degree of DOCTOR

to be defended in public on

Wednesday, 6 February 2013, at 10:00 hours

In Delft, the Netherlands

by

Collin Calvin MABIZA

born in Chivhu, Zimbabwe

Bachelor of Arts Honours in Geography and Environmental Science,
Masters in Environmental Policy and Planning, University of Zimbabwe

This dissertation has been approved by the supervisor:
Prof. dr. ir. P. van der Zaag

Composition of the Doctoral Committee:

Chairman	Rector Magnificus TU Delft
Vice-Chairman	Rector UNESCO-IHE
Prof. dr. ir. P. van der Zaag	UNESCO-IHE / Delft University of Technology, Supervisor
Prof. dr. ir. H.H.G. Savenije	Delft University of Technology
Prof. dr. ir. C.M.S. de Fraiture	UNESCO-IHE / Wageningen University
Prof. dr. D.S. Tevera	University of Swaziland, Matsapha, Swaziland
Dr. E. Manzungu	University of Zimbabwe, Harare, Zimbabwe
Dr. ir. J.A. Bolding	Wageningen University
Prof. dr. ir. N.C. van de Giesen	Delft University of Technology, reserve member

CRC Press/Balkema is an imprint of the Taylor & Francis Group, an informa business

© 2013, C.C.Mabiza

Published by:
CRC Press/Balkema
PO Box 11320, 2301 EH Leiden, The Netherlands
e-mail: Pub.NL@taylorandfrancis.com
www.crcpress.com - www.taylorandfrancis.com

ISBN: 978-1-138-00036-0 (Taylor & Francis Group)

Abstract

Most of southern Africa is semi-arid. Parts of the region, such as the Limpopo river basin, are characterised by low rainfall totals. More than half of the region's population has limited access to water. High dependence on rainfed agriculture to a large extent accounts for food insecurity and high incidence of poverty in the Limpopo river basin. These factors make improved water resources management a critical need as it can potentially contribute towards raising the standards of human welfare and socio-economic development. As part of efforts towards improving water resources management most countries riparian to the basin have adopted Integrated Water Resources Management (IWRM) as a framework within which water is managed. Early adopters of IWRM, such as Zimbabwe, have already gone past the first decade of implementing IWRM. Given the time that has passed since the adoption of IWRM, it is important that an analysis be made on whether, and how, IWRM has improved practices in water resources management and contributed towards improved livelihoods within the river basin. This is critical, either for the improvement of IWRM as it is being implemented, or for the development of new water management frameworks.

Using a bottom-up approach, this study analyses water management practices and livelihoods at the local level. The context of the study is a river basin under stress, both in terms of the agro-ecological (natural) conditions, and in terms of socio-politico-economic conditions. The purpose of the study is to try and understand *what are the practices in water resources management at the local level, and what are the drivers of those practices?* Water resources management, among other things, is supposed to improve livelihoods, and this need to understand livelihoods explains why a bottom-up approach was chosen for this study. The study also opts for a multi-foci approach to broaden understanding of practices in water resources management in different livelihood contexts. In literature it is common to find analyses that fragment livelihood issues, such as focusing on access to water for domestic uses only, or on water for productive uses only. This gives an incomplete picture of how water resources are managed at the local level. The study adopts a case study approach, and analyses five cases on: practices in access to water for domestic and productive uses, efforts at sustaining livelihoods and the environment, water management for agriculture, contestations over urban water services and river basin planning. In all cases an understanding of what actually drives practices in water resources management was sought.

The study made a review of IWRM as a water management framework. The review analysed IWRM from a conceptual perspective and from a practical standpoint in terms of implementing the framework. At the conceptual level a critical observation made was that IWRM appears not to directly target improving livelihoods. This to a large extent explains why the development of physical infrastructure, which can improve people's access to water, is often not considered in IWRM programmes. Furthermore, it was observed that IWRM is not clear on what needs to be integrated. On the aspect of integration IWRM was found to be under criticism on two fronts. On the one hand IWRM is accused of not being realistic about what can be integrated. It is considered to be bringing too many issues under the umbrella of water resources management. On the other hand it is attacked for not being integrative enough. Critics argue that a lot of important issues are left out of the framework. However, challenges encountered in the implementation of IWRM have led to calls for 'Light IWRM' to be substituted for full IWRM.

The first empirical case presented in this study tried to answer the question, *what drives practices in water resources management at the local level?* Practices of water users were analysed at different sources of water and water infrastructure, specifically a borehole, a wetland and a wind-powered water infrastructure. The concepts of practice, interaction and institutional bricolage were used to investigate local water resources management. Focus group discussions, interviews and participant observer methods were used to gather data for the chapter. The study found out that, although in Zimbabwe IWRM has been in place for about a decade, practices in water resources management at the local level were still taking place outside the framework. Catchment and subcatchment councils were found to be absent at the local level, and therefore not influencing practices in water resources management. The chapter found that practices in water resources management at the local level were mainly influenced by the socio-economic and physical context in which water exists and is managed. At some waterpoints it was found that the semi-arid conditions in the catchment, combined with dysfunctional technology, heightened perceptions of an impending water scarcity. This to an extent account for why rules put in place by local committees were often broken by water users. An important finding was that at waterpoints where water was used for productive uses, infrastructure tended to be better maintained than where it was used only for domestic uses. This suggests that within the IWRM framework more effort should be made to ensure that vulnerable groups in society have increased access to water for productive uses. The chapter concluded that the approach of forming new institutions in response to water resources management challenges needs to be questioned. There is potential that water resources management can be improved by addressing livelihood concerns, such as through providing water for productive uses.

An investigation into how local actors try to 'sustain' livelihoods and the environment (catchment management) was also made, as the second case of the thesis. The context in which the analysis was made was characterised by an adverse socio-economic and physical environment. Analysis centred on two cases, one on gold panning and another on a gully reclamation project, both activities which have a bearing on water resources. It was found that actors at the local level were being driven by the socio-economic and semi-arid environment to exploit the physical and the institutional resources at their disposal. Actors were found to adopt contradictory practices, on the one hand engaging in environmentally-friendly projects, and on the other hand creating environmental hazards in the same catchment. Taking part in environmental reclamation projects implemented by non-state actors as Food For Work (FFW) projects enabled local actors to access food hand-outs. However, when non-state actors stopped handing out food actors dropped out of the environmental management project. This supports the view that although FFW try to address livelihood challenges, the reality is that the majority of the world's vulnerable population are affected by chronic hunger which cannot be solved through piecemeal and short-term efforts. Furthermore, treating water as an economic good as currently formulated in the fourth Dublin Principle can potentially worsen environmental management by reducing access to water by the poor. An improvement of that particular Dublin Principle is therefore necessary.

Another aspect of local water management investigated was that of agricultural water management at the field-level, focusing on rain-fed farming, which made the third case. Narratives of different actors were used to analyse how field level water management techniques are being promoted as conservation agriculture among smallholder farmers. It was found that the main tenets of conservation agriculture being promoted include minimum tillage, nutrient management and mulching. This package of practices was found to be beyond the capacity of smallholders, the major complaint being that the labour demands of the practices are too high. However, smallholder farmers were found to be adopting conservation agriculture because of offers of free seed and other inputs from non-state actors. A conclusion reached in the chapter was that smallholders were aligning themselves to non-state actors promoting conservation agriculture for livelihood purposes, and not necessarily because of the potential of conservation agriculture to increase yields.

The fourth case made an analysis of how context accounts for the contestations over the control of urban water services. The case was based on the proposal by central government for the national water authority to take-over the city of Bulawayo's water services. This resulted in contestation over the city's water services between the local authority and the city's residents on one side, and central government on the other. It was found that central government advanced reasons of efficiency to justify the proposed takeover. However, at the heart of the contestation was the symbolic meaning of water. To the city residents water services represented defiance, to nature as the city is in a semi-arid environment, and to central government as the city had built its own dams over the years. To the government capturing the water services would allow it to more directly control the urban authority. A conclusion reached was that, contrary to prescriptions of international financial institutions, economic efficiency alone cannot explain why certain water resources management models are accepted while others are rejected. In some cases water resources management models are accepted or rejected based on the social value attached to water. This finding suggests that the manner in which IWRM has to date been promoted and subsequently been accepted or rejected in the developing world could possibly be linked to governance in general rather than to water specific issues.

The last empirical chapter analysed processes in river basin planning, and whether or not their outcomes match the livelihood realities of water users. This was done by analysing processes through which the draft catchment outline plan for the Mzingwane catchment was made. It was found that the making of the draft plan involved minimal participation by stakeholders as the strategies for collecting data for the plan failed to garner the support of stakeholders. The socio-economic context in which the draft plan was made was a major hindrance to the involvement of stakeholders. Thus although guidelines and the law provided for consultations, these could not be effectively held. The limited participation of stakeholders explains why the draft plan failed to address livelihood challenges, such as those of access to water for both domestic and productive uses.

An overall conclusion which the study makes is that, although the adoption of IWRM assumes that water resources management can be improved through internationally acclaimed frameworks, local socio-politico-economic and physical factors to a large extent drive the water resources management practice. Thus improving water resources management cannot be tackled in isolation from improving livelihoods.

Dedication

Blessed assurance, Jesus is mine!
O what a foretaste of glory divine!
Heir of salvation, purchase of God,
Born of his Spirit, washed in his blood.

This is my story, this is my song,
praising my Savior all the day long;
this is my story, this is my song,
praising my Savior all the day long.

Perfect submission, perfect delight,
visions of rapture now burst on my sight;
angels descending bring from above
echoes of mercy, whispers of love.

Perfect submission, all is at rest;
I in my Savior am happy and blest,
watching and waiting, looking above,
filled with his goodness, lost in his love.

Fanny J. Crosby, 1820-1915

Acknowledgement

This thesis would not have been possible without the guidance of Prof. Van der Zaag, my promoter, and Dr. Manzungu my supervisor, doyens of water resources management. When this PhD began many told me how lucky I was to be supervised by these two eminent figures in water resources management, and now I feel humbled I was their student, which is why the thesis is dedicated to them. Indeed this thesis reflects their wisdom, diligence, hard work and deep understanding of water resources management issues in southern Africa in particular and the world over in general. May they continue to be the guiding lights in the water sector, and the world continue to benefit from their expertise. The generous financial support of the Challenge Programme on Water and Food (CPWF) (PN 17) made this research possible. WaterNet and the International Foundation for Science (IFS) also contributed financially to this study, and their help is appreciated. The Department of Civil Engineering of the University of Zimbabwe offered me the position of Research Associate which facilitated this research, and I am grateful for that. In particular I would like to thank the past Chairmen of DCE, Dr. Salahuddin and Eng. Hoko, and the current Chairman, Dr. Makurira, for their unwavering support to my project. Prof. Nhapi and Dr. Tumbare thank you for your support. I would also like to extent my gratitude to Mrs. Sadazi, Mrs. Musiniwa and Ms. Chivhinge who were always ready to help me.

Most of the fieldwork for this study was carried out in Ward 1, Insiza District, Zimbabwe, and I would like to thank the people of that community for welcoming me. I am particularly grateful to MaTshili of Thandanani Village, and Mr. Mpofu of Mpumelelo Village. From these seasoned teachers I learnt a lot about Ward. The staff at Tshazi Secondary School gave me a place to stay, and for that I am very grateful. Khumbulani Ndlovu helped me with data collection, I am grateful for the assistance. I am also indebted to ZINWA staff at Head Office and of the Mzingwane Branch who gave me access to some of their records.

Through the Challenge Programme (PN 17) I met David Love, Manu Magombeyi, Paiva Manguambe, Alexander Mhizha, Walter Mupangwa and Pinimidzai Sithole, fellow PhD researchers. Thank you gentlemen for being such good friends. I have since forgiven all of you for the conspiracy to poison me in Johannesburg, South Africa. I would also like to thank WaterNet staff, in particular Dr. Gumbo, Mrs. Hondo, Ms. Makopa-who assisted me in many ways.

I am indebted to my Dutch friends, staff and colleagues at IHE. Dr. Ahlers played a big role in guiding my research, especially in first few years, thank you very much. Dr. Marloes Mul thank you for translating my abstract into Dutch. I will always remember your kindness -I lost count of the number of meals and braais we had at your place, both in Harare and in Delft. Thanks Marloes. Thanks also to Jolanda Booths and Jacqui for organizing my trips to Delft and for making my stay in the Netherlands pleasant and comfortable. In the Department of Water Management and Institutions I met PhD fellows who became friends- Hans, Eric, Xiao, Ilyas, among others. It was nice being in your company. Hilda and Harold, my Dutch friends-thank you for opening your home to me. Through you I learnt about the Dutch society, how the Dutch live, farm and celebrate weddings! I will always cherish your friendship. I would also like to extent my gratitude to fellow Christians at IREF. The church services really uplifted me. Back home I would like to thank my friends and brethren in the Reformed Church in Zimbabwe-thank you for praying with and for me. Last but not least I would like to thank my family for their love and support. My parents prayed for me without ceasing, and I am eternally grateful for that. I want to be a praying parent because of you. My sisters proved to be my all-weather friends, I can never thank you enough for that. Connie, Tino, Munesuishe and Kayla, thank you for being a wonderful bunch. You provided comic relief when I needed it most. Mumu, I must have been poor company for you to remind me that I only get a goodnight hug when I come to bed (early). Finally, I would like to thank Nana, my darling sweetheart. Thanks for being patient with me and for encouraging me to soldier on. Each time I try to thank you, or say I love you, I realize how few the words I know are and am tempted to go back to school! I love you Nana, and always will.

Table of Contents

Chapter 3

Integrated Water Resources Management and livelihoods: do they meet?   29

Chapter 4

Dynamics and complexities of practices in local water management in Zimbabwe in the IWRM era   45

Chapter 5

Livelihood strategies and environmental management in Zhulube micro-catchment   67

Chapter 6

Sowing seeds of hope: the case of conservation agriculture in the smallholder farming sector   **95**

Chapter 7

Contestations and coalitions in urban water supply: the state, the city and the politics of water in Bulawayo, Zimbabwe   **115**

Chapter 8

Processes in river basin planning: the case of the Limpopo river basin in Zimbabwe   **133**

List of Figures

List of Tables

List of boxes

List of acronyms

ADP	Area Development Programme
AGRITEX	Agriculture Research and Extension Services
AREX	Agriculture Research Extension
BCC	Bulawayo City Council
BEAM	Basic Education Assistance Module
CAMPFIRE	Communal Area Management Programme For Indigenous Resources
CBNRM	Community Based Natural Resources Management
CC	Catchment Council
CGIAR	Consultative Group on International Agricultural Research
CMA	Catchment Management Agency
CPWF	Challenge Programme on Water and Food
DDF	District Development Fund
DFI	Direct Foreign Investment
DFID	Department for International Development
DNR	Department of Natural Resources
DRSS	Department of Research and Extension Services
DWA	Department of Water Affairs
DWAF	Department of Water and Forestry Affairs
DWD	Department of Water Development
EIA	Environmental Impact Assessment
EMA	Environmental Management Agency
EMP	Environmental Management Plan
ESAP	Economic Structural Adjustment Programme
FAO	Food and Agriculture Organisation
FTLRP	Fast Track Land Reform Programme
GDP	Gross Domestic Product
GMB	Grain Marketing Board
GTZ	German Organisation for Technical Cooperation (Gesellschaft für Technische Zusammenarbeit)
GWP	Global Water Partnership
ICMA	Inkomati Catchment Management Agency
ICP	Integrated Catchment Planning
ICRISAT	International Crops Research Institute for the Semi-Arid Tropics
INRM	Integrated Natural Resources Management
IRWSSP	Integrated Rural Water Supply and Sanitation Programme
ITCZ	Inter-tropical Convergence Zone
IWRM	Integrated Water Resources Management
LGPO	Local Government Promotional Officer
LIMCOM	Limpopo Watercourse Commission
MDGs	Millennium Development Goals
MRRWD	Ministry of Water and Rural Resources Development

MUS	Multiple Use water Services
NGO	Non-governmental organization
NLHA	Native Land Husbandry Act
NWRS	National Water Resources Strategy
PASS	Poverty Assessment Study Survey
PRP	Protracted Relief Programme
RBDAs	River Basin Development Authorities
RBO	River Basin Organisation
RBZ	Reserve Bank of Zimbabwe
RDC	Rural District Council
RDDC	Rural District Development Committee
RDDP	Rural District Development Plan
SADC	Southern Africa Development Community
SCC	Subcatchment Council
TTL	Tribal Trust Lands
TVA	Tennessee Valley Authority
UNDP	United Nations Development Programme
UNICEF	United Nations Children's Fund
VIDCO	Village Development Committee
WADCO	Ward Development Committee
WCED	World Commission on Environment and Development
WFP	World Food Programme
WPC	Waterpoint Committee
WRC	Water Research Commission
WSSD	World Summit on Sustainable Development
WUA	Water Users Association
WW2	World War Two
ZANU (P.F)	Zimbabwe African National Union (Patriotic Front)
ZAPU (P.F)	Zimbabwe African People's Union (Patriotic Front)
ZIMVAC	Zimbabwe Vulnerability Assessment Committee
ZINWA	Zimbabwe National Water Authority

Chapter 1

Introduction

1.1 Background: local resources management in Zimbabwe

This study is about local water management practices and processes in a semi-arid area, specifically in southern Africa's Limpopo river basin. Socio-economic development in southern Africa has historically been, and continues to be closely tied to water resources management (Swatuk, 2010). This is to a large extent true for Zimbabwe, which is one of the countries riparian to the Limpopo river basin. Successive governments in the country have made water resources management an important aspect of state formation processes (Scott, 1998). Although the colonisation of the country in the 1890s was largely on the lure of mineral wealth, diamonds in particular, it was water which ended up as the most important resource in the colony. This was so because the mineral deposits discovered in the country turned out to be less than had been anticipated by the settlers. Disappointed colonisers turned to agriculture as an economic activity, which led to the rise of irrigated agriculture in the country (Campbell, 2003; Manzungu and Machiridza, 2009). Consequently water resources management became the lifeblood of the economy and a source of power (Campbell, 2003, Swatuk, 2010). Arguably, from then on water has been an important determinant of well-being, be it of the economy or of the different actors in the country.

The necessity of water to life, and its importance to socio-economic development explains man's perennial quest to improving water resources management. Arguably, in the wake of the Rio Earth Summit of 1992, the challenge for improved water resources management has been further expanded to include management of water for environmentally sustainability. This quest for improved water resources management is evident in the number of paradigms through which water resources management has gone (Allan, 2003). Among the paradigms is the hydraulic mission, which was characterised by the development of huge infrastructure such as dams. These were meant to harness water to power industry-based economies (*ibid*). However, it is important to note that in the field of natural resources management in general, institutional solutions have often been considered to be the antidote to the development conundrum. In both pre-independent and independent Zimbabwe such solutions have defined the dynamics between central government and resource users. Through pieces of legislation such as the Natural Resources Forest Produce Act (1928) and the Natural Resources Act (1941) and the Native Land Husbandry Act (1952) the colonial government centralised natural resources management (McGregor, 1995; Mandondo, 2000). Legislation gave government agents the power to intervene at the local level to enforce resource conservation, for example, through forced soil conservation works, destocking and prohibition of cultivation within a certain distance from stream-banks. This approach to

1

natural resources management created tension between the government and local actors. This was because natural resource conservation jeopardised the livelihoods of Africans. Destocking, for example, threatened to wipe out the herd owned by Africans. Livestock are critical livelihood assets among Africans. It is important to note that agents of the government, such as extension workers, became part of the machinery enforcing natural resources management measures (Bolding, 2004). Extension workers forced smallholder farmers to construct soil conservation works, contour ridges being the most common structures and most unpopular among the African farmers (*ibid*). Unfortunately solutions that were suggested for dealing with resource management challenges were based on a limited understanding of the complexity of the African socio-economy. Often linear relationships between resource management problems and livelihood strategies of Africans were erroneously drawn. Forced destocking and contour-ridging are examples of solutions that were based on limited analyses of the root cause of challenges in resource management. This to a large extent explains why resisting resource conservation measures came to be part of the struggle for independence (McGregor, 1995). In the light of these observations it is critical to analyse if approaches to resource management being implemented in the modern day take into account the complexity of livelihoods.

Upon attaining independence in 1980, the new government took steps to create a balance between natural resources management and the improvement of livelihoods. The government's approach towards sustainable natural resources management focused on increasing participation and spreading the benefits of resources management to communities. This shift was encouraged by global trends which promoted community based natural resources management (CBNRM) as the most sustainable way of managing natural resources. Through the Communal Areas Management Programme for Indigenous Resources (CAMPFIRE), for example, the government devolved the management of wildlife to Rural District Councils (RDCs) (Madzudzo and Hawkes, 1996; Mapedza, 2007). This programme enabled communities in areas bordering national parks and other areas with high wildlife populations to benefit from income derived from tourism, for example. The impact of such programmes on livelihoods was direct as communities were able to use such income to develop water infrastructure and educational facilities, for example. In return communities were supposed to contribute towards the protection of wildlife. This win-win scenario to an extent made resource management sustainable as it brought tangible livelihood benefits to households. Unfortunately, by the late 1990s the programme was inflicted with challenges associated with a declining economy and deteriorating political environment (Mapedza, 2007).

While the above has sketched how resource management in general has evolved over time, the focus of this thesis is on local water management, especially given the importance of water in the semi-arid Limpopo river basin. The next section focuses on how water resources management has evolved in Zimbabwe.

2

1.2 Local water resources management: past to present

Over the years, in both the pre-independence and independence era, state and non-state actors have intervened in water resources management in smallholder farming areas. A dichotomy of interventions is discernible, those directed at water for productive uses, and those with a focus on water for domestic uses. Missionaries, who were among the first white settlers in the country, set up irrigation schemes as far back as 1912 (Zawe, 2000). This approach was continued by successive governments. Such interventions were part of a strategy meant to increase productivity in the smallholder sector. Unfortunately operational and policy-related factors account for limited contribution of irrigation towards the improvement of livelihoods. A weakness of the colonial government's policy was that it placed emphasis on productive water uses without due regard to domestic water needs. This omission was puzzling given that although policy prescriptions purported to address food security, domestic water, which is a critical livelihood need, was not given due consideration.

As far as domestic water was concerned, the Zimbabwean government focused on increasing access to water by the rural majority. Programmes such as the Integrated Rural Water Supply and Sanitation Programme (IRWSSP) through which boreholes and deep wells were sunk in the rural areas were initiated to improve rural livelihoods (Katsi *et al.*, 2007). However, shortage of spares and skilled personnel to repair and maintain the water infrastructure resulted in little changes in clean water availability at the local level (Cleaver, 1990). The Zimbabwean government did not only address the infrastructural dimension of water resources management, more importantly it created a tiered-institutional structure to try and address the challenges in the water sector (Cleaver, 1990). A critical component of the new institutional structure were the Waterpoint Committees (WPCs). Each borehole provided was supposed to have a WPC to oversee the use, repair and maintenance of the infrastructure. One can suggest that in independent Zimbabwe this marked the first major institutional solution prescribed for the water sector. What remains unclear is how these committees were supposed to interface with traditional institutions already in place at the community level. Although the WPCs can be considered to be an innovation, the approach taken by the government marked a continuation of a top-to-bottom approach to solving resource management problem. Such approaches were common in pre-independent Zimbabwe.

Interventions highlighted above took place before the country adopted IWRM. With the encouragement of international donors, and in response to local realities, Zimbabwe, together with several other southern African countries, embraced IWRM in an attempt to redress injustices of inequity in access to water. The framework was promoted by international actors such as the Global Water Partnership (GWP) on the argument that implementing it yields socio-economic and ecological benefits. Subsequent to the adoption of IWRM Zimbabwe promulgated a new water law and policy in 1998. However, this adoption of a new water management framework, which can be considered as an intervention in the water sector, does not appear to have had much impact on water resources management at the local level. Among the challenges observed is that of the disconnect between the water users and river basin organisations, and also between the river basin organisations at various hydrological scales (Fatch *et al.*, 2010).

It appears that the various institutional interventions in water resources management initiated by both state and non-state actors have not been based on a bottom-up perspective. As was highlighted above, some of the interventions have been based on a narrow or compartmentalised perspective of water resources management. An example is that of cases whereby projects for providing water for productive uses have not been coupled with domestic water supply projects. This can be attributed to a limited understanding of the link between water and rural livelihoods, in particular how water forms the basis of rural livelihoods and draws different actors and issues together. Furthermore, interventions, by nature, are based on the assumption by agents of development, be they state or non-state actors, that local practices and conditions can be improved (Long and Van der Ploeg, 1989). However, in practice interventions are not painted on a blank canvas. They have to deal with existing realities, be they socio-economic or those relating to the physical environment. When analysing interventions in water resources management one can ask, *what new institutions have been created, how are they working*? Closely related, it can also be asked, how do local actors take advantage of interventions so as to 'secure' their livelihoods? Notably, 'security' when applied to household needs, such as food, is often used to denote a situation whereby access to food is long term and in the right nutritional quantities and meets personal choices (FAO, 1996). Interventions in water resources management also do not seem to value the dynamics and practices in water resources management at the local level, and how these are shaped by, and in turn shape livelihoods.

Interestingly, with the advent of IWRM and the claims that the framework is able to facilitate socio-economic development, the local level has now become subject to multiple interventions in the name of improved water resources management for improved livelihoods. Interventions in the management of water infrastructure, management of water at the field level for crop production, and even the so-called watershed management projects have now become part and parcel of water resources management. Water resources management in practice has therefore become an effort to align the practices of different actors operating at different institutional levels and spatial scales to achieve one articulate goal, which is that of improved livelihoods. This calls for analysis with multiple foci to understand how the relationship between the multiple water resources management strands and livelihoods.

1.3 Problem statement

It has been suggested that improved water resources management can potentially contribute towards improving rural livelihoods (Merrey *et al.*, 2005). However, what is not clear is, what does improved water resources management entail? How does it, or can it interface with local realities, and how can it contribute towards improved livelihoods? Arguably, improving livelihoods through water resources management has remained an elusive goal to date because, among other things, the water resources management-livelihoods nexus is not clearly understood. Such an understanding might be possible if a bottom-up and multi-disciplinary inquiry to the nexus is taken. Literature suggests that such an approach has not been made to date. Common approaches have, for example, analysed institutions in isolation, such as what Sithole (2001), Latham (2002), Kujinga and Jonker (2006), Merrey (2007), Chikozho (2008) and Fatch et al., (2010) have done. In some cases, water resources management at the field level has been tackled without adequate attention being paid to livelihoods (see, for example, Hove and Twomlow (2006) and Mupangwa et al.,(2008). Thus an urgent need is to study water resources management and livelihoods from multiple perspectives.

Advocates of IWRM, such as the GWP, suggest that implementing the framework delivers socio-economic and ecological benefits. However, in the light of such claims it is necessary to pause and ask, can livelihoods be improved within the IWRM framework? Furthermore it can also be asked, is IWRM able to combine with local practices and contribute towards improving livelihoods? In other words, upon encountering drivers of processes at the local level, what becomes of water resources management? A critical point to analyse in this regard is whether enough is known about local practices in water resources management to enable IWRM to be applied at the local level. If IWRM is to be better formulated, implemented and work for local water users it is imperative to investigate how practices in local water management emerge, what shapes them, and how they fit and reflect local (and other) realities and livelihoods. Closely related is that, since rural livelihoods in southern Africa are mostly agro-based, how does, or how can IWRM contribute towards improving agricultural productivity and in that way improve livelihoods? Issues like the ones raised above have implications on the practices of organisations in water resources management. Zimbabwe, for example, in line with IWRM Principles, has created river basin organisations at the subcatchment and catchment level. These organisations can be seen as forming the bridge between the theory and the actual practice of water resources management. The question in search for an answer is, how do these organisations translate policy into action, and how does that contribute towards improving livelihoods? This study therefore seeks to make a bottom-up investigation into practices and interventions in water resources management at the local level, and how these how these impact on livelihoods.

1.4 Research significance and justification

Although substantial research on water resources management in Zimbabwe has been made, most of the studies cover the period before the country experienced a severe economic crisis. The economic crisis was characterised by high inflation rates and reached its peak around 2008 when annual inflation was estimated to be in the millions. Accompanying the crisis was a general shortage of commodities, agricultural inputs included. The combined effect of the economic crisis and natural factors such as low rainfall totals exacerbated poverty especially among the rural populace. These observations make a study on water resources management at the local level, which is where livelihoods are most endangered, a priority. Although Zimbabwe enacted IWRM-based water law and policy in 1998, the framework continues to guide water resources management in the country, which makes it necessary to examine whether or not the framework fits the context in which it is being applied. It is important that the case for whether to maintain, modify or outright reject current IWRM-based water resources management practices be based on empirical evidence. It is hoped that insights from this study will contribute towards that endeavour.

This research was carried out in the context of the Challenge Programme on Water and Food (CPWF) Project Number 17 (PN 17) entitled, 'The Challenge of Integrated Water Resources Management for Improved Rural Livelihoods: Managing Risk, Mitigating Drought and Improving Water Productivity in the Water Scarce Limpopo Basin.' The CPWF is a trans-institutional inter-disciplinary project funded by the Consultative Group on International Agricultural Research (CGIAR). The main objective of the CPWF PN 17 was to find possible solutions to the challenges of water and food at the local level.

1.5 Research questions

The purpose of this research is to enhance the understanding of the water resources management-livelihoods nexus. The main research question guiding this study is: *what are the drivers influencing practices in water resources management and livelihoods in the Limpopo's Mzingwane catchment?* The following are the specific research questions that will guide the study:

1. *What drives practices in water resources management at the local level?*
2. *How do local actors try to 'sustain' livelihoods and catchment management?*
3. *What field-level practices are being promoted to manage water in smallholder farming, and how can their adoption or rejection be explained?*
4. *How does context help in understanding contestations over the control of urban water services?*
5. *What accounts for processes through which river basin plans are made, and the outcomes of those processes?*

1.6 Theoretical framework and concepts used in the study

1.6.1 Theoretical framework

This study attempts to combine the theory of social construction with the concept of water control. This theory is primarily concerned with the way in which actors structure experience and analyse the world (Owen, 1995). It has its foundation in the area of (social) psychology, and its emergence was based on the argument for social processes as origins of knowledge and practices (Gergen, 1985). Social construction inquiry concerned itself with the processes by which actors came to describe, explain or account for the world they live in (*ibid*). Proponents of the theory argue that actors are influenced by social and interpersonal influences, thus actors' discourse about the world is an artifact of communal interchange (Gergen, 1985). It places emphasis on the complexity and inter-relatedness of the many facets of individuals within their communities (Cromby and Nightingale, 1999). In this thesis the theory of social construction will be used to try and understand how practices in water resources emerge, in particular the inter-relatedness between practices in water resources management and the livelihoods context in which it takes place.

However, applying this theory to water resources management has the major shortfall in that practices in water resources management are more than outcomes of social processes. For purposes of this study therefore, the theory of social construction will be combined with the concept of water control, which views the constituent elements of water resources management as being: (i) the technical/physical elements, (ii) the organisational components, and (iii) a socio-economic and regulatory dimensions (Mollinga, 2008). Taking such a theoretical standpoint potentially enables one to analyse water resources management and livelihoods as they are determined by the context within which the actors live. This context comprises the physical environment which determines the natural resources, including water as a physical resource, available to the actors. However, man is not a passive actor in the equation but plays a significant role in influencing the material dimension, which also includes the water infrastructure through which households access water. Thus an understanding of water resources management and livelihoods becomes possible by alluding to both natural and human elements which make up the system. This is the framework through which this study is conducted.

Livelihoods defined
The key term used in this study is that of livelihoods. Analyses of rural livelihoods was, until the 1990s, largely based on quantifiable socio-economic factors, such as income, cost of labour and related variables. Such variables could not only be quantified, but importantly a monetary value could also be attached to them (Chambers, 2004). However, it became increasingly clear that rural livelihoods were complex and could not be reduced to only what could be converted to a monetary value. Access to resources, for example, was noted to be key in determining the quality of livelihoods, yet this was often left out in analyses of rural livelihoods. Such observations led to the rise of the livelihoods framework which gave attention to the different facets of rural livelihoods. Chambers (2004) defines a livelihood as being adequate stocks and flows of food and cash to meet basic needs and to support well-being. A sustainable livelihood comprises people, their capabilities and their means of living, including food, income and assets (Chambers and Conway, 1991). Tangible assets can be resources, while intangible assets can be claims and access. A livelihood is sustainable when it can cope with and recover from stresses and shocks and provide for future generations. In the communal areas it is important that livelihoods become sustainable to ensure the wellbeing of people. Sustainable livelihoods also contribute towards ensuring that environmental stocks are not depleted. In this thesis livelihoods will be considered in the context of how people ensure a living.

1.6.2 Concepts used in the study
Barnett (2001:6) defines a concept as a general notion or idea with a particular (but potentially contested) meaning. Concepts help to explain, classify and organise thoughts (*ibid*). The overarching concepts used in this thesis are that of livelihoods and institutions. In this thesis water resources management is taken to be the rules governing

access to, and use of the resource, and for that reason the concept of institutions is central to the thesis. However, apart from this overarching concept, each chapter also makes use of concepts which apply to specific cases under investigation. Such concepts are meant to give a clearer understanding of institutions and water resources management. Here the following concepts are discussed briefly: livelihoods, institutions, practice, scale, and intervention.

Institutions

Institutions are associated with ideal behaviour and expectations and which can be used as a generic concept for the variety of rules that help to pattern behaviour, norms, folkways, mores, customs, convention and law (Coward, 1980). Institutional arrangements are the specific rules that individuals use to relate to each other (Ostrom *et al.*, 1993). Institutions have also been defined simply as 'rules of the game' (North, 1990). Institutions can constrain or enable behaviour (Bromley, 1989). They make natural resources management and other aspects of life orderly and to an extent more predictable. Ostrom (1990), arguing from the new institutional perspective, sees institutions as being outcomes of rational crafting. This view is opposed by Cleaver (2000) who sees institutions as being a result of multiple processes, conscious and unconscious acts which are influenced by acceptable patterns of behaviour and social relationships. She suggests that institutions exist in a variety of forms and are constantly evolving. In water resources management institutions play the vital role of setting the boundary for, among other things, use, access, sharing and conservation of water resources. The special nature of water necessitates effective institutions. In this thesis institutions will be taken as the rules which regulate resource use and management. The thesis uses the theory of institutional bricolage as put forward by Cleaver (2000) which suggests that rules are emergent, influenced by the physical environment in which the resource occurs and is used, and also by the cultural beliefs and values of the resource users. Since rules are abstract, the thesis also argues that institutions can be studied empirically by analysing practices of water users, for example, at waterpoints as they access water.

Practice

Institutions as rules cannot be observed, they are abstract, however, their effect on water users manifests in what water users do, which is observable. This presents the challenge of how to investigate institutions in water resources management. To solve that problem, the study analyses practices. Practice is *'routinized' type of behaviour which consists of several elements, interconnected to one another: forms of bodily activities, forms of mental activities, 'things' and their use, a background knowledge in the form of understanding, know-how, states of emotion and motivational knowledge,'* (Reckwitz (2002:249). A practice is social since it is a type of behaving and understanding that appears at different locales and at different points of time (*ibid*). The concept is useful for, among other things, analysing how actions of people are related to material objects (Van der Zaag, 1992). In this thesis the concept of practice will be used to analyse how water resources are managed at the local level. Moreover, aspects such as catchment planning can be traced in terms of how they are made through observing practices of catchment councils. Therefore practice in this thesis will not be limited to actors with agency (human beings) but will also apply to institutional arrangements which can be considered to get and express their agency through actors working in them. A point to note that water is used and managed at different spatial scales which necessitates that the concept of scale be analysed.

Scale

Scale is the spatial, temporal, quantitative or analytic dimensions used to measure and study any phenomenon (Gibson *et al.*, 2000). Scale can be a social construction (Lebel *et al.*, 2005). Closely associated with the concept of scale are the issues of claims (Kurtz, 2003). Typically resource users determine who can or cannot have access to a resource by looking at scale issues among other things. Although scale has always been important in natural resources management in general, the ascension of IWRM and its insistence that water should be managed along hydrological boundaries makes it necessary for scale issues to be analysed closely in the water sector. This is so because although within the IWRM framework water is managed along hydrological boundaries, in reality water users live in socially and politically constructed boundaries. Thus water resources management can be argued to be an attempt to align different scales so that socio-economic and ecological benefits can be derived. Furthermore, water resources management can also be considered as an effort to align the practices and aims of actors operating at different scales. In this thesis the concept of scale will be used to analyse access to water and management of water infrastructure. By problematising the concept of scale, the thesis will show that actors operating at different scales have different water resources management concerns which makes simple solutions to challenges in the sector elusive. The multiplicity of water resources management concerns will be shown to occur at the field scale, micro-catchment level, and the catchment level.

Narratives

Narratives have been defined as a storyline that gives an interpretation of some phenomena (Molle, 2008). They are often used in the social sciences as a way of making social reality accessible through words and stories (Hyden, 2008). Narratives enable social actors to express their understanding of reality (Rimmon-Kenan, 2006). They can be collected through surveys, interviews or observations. This makes them to be perceived as being subjective and not objective, and as not being fact (*ibid*). This perception does not acknowledge the fact that even policy makers and other agents of development design interventions based on their own interpretation of reality, which is also subjective. Thus contrary to common perceptions, narratives are often key in the design of interventions. Thus it is only reasonable that policy evaluation must also take into account narratives of local actors affected by the intervention. In this thesis narratives will be used to capture the perceptions of local actors, and to understand the power dynamics and ideologies which have been used to promote interventions in local water management.

1.7 Research design and data collection methods

This study purposefully sought to investigate practices in water resources management in a particular context within a specific setting. The basis of this approach was that water resources management has to address specific needs of a specific group of water users within a specific area. The research was therefore designed in such a way that an in-depth study of a given context of water users, uses and practices could be done. For this reason the study used the case study approach. Data collection methods which were selected to carry out the study were chosen on the basis of their capacity to enable an in-depth study of the cases selected.

1.7.1 The case study approach

The study employed the case study approach. Yin (2009) describes the case study method as an empirical inquiry that investigates a contemporary phenomenon in depth and within its real-life context. This research was interested in understanding practices in water resources management, exploring the relation between livelihoods and the context within which the livelihoods are situated. The survey method, for example, was not chosen because one of its limitations is that it attains validity through statistical generalization, which is not what this study wanted to do. This study assumes context to be an important factor in water resources management, so to understand the multiple-realities of water resources management case studies were considered to be the best approach. This study uses multiple cases to bring out a detailed understanding of the interface between water resources management and livelihoods, and between livelihoods and institutions. Multiple cases make it possible for conclusions to be made across the cases (*ibid*). Scientific validity when using such an approach is obtained, at the methodological level, through the use of multiple sources of evidence, and at the analysis level, through theoretical propositions.

Case selection

In this study five cases were selected: (1) on water management at the local level, (2) on environmental management and livelihoods at the local level, (3) on conservation farming, (4) on contestations over the management of urban water services, and (5) on river basin planning. The selection of cases was based on the need to understand the water resources management-livelihoods relationship from different contexts. Literature shows often research in water resources management focuses on a single issue, such as smallholder irrigation, or stakeholder participation in catchment councils. This single-issue approach does not adequately capture how, in some cases, different water resources management issues can affect livelihoods of more or less the same actors. In this study an effort was made to broaden the understanding of the water resources management-livelihoods relationship by analysing: (a) issues that affect both urban and rural water users, (b) different issues that different categories of water users experience within a micro-catchment, and (c) challenges that formal institutions managing water face as they try to address livelihood issues of water users. In the rural areas the management of water for domestic uses, and of water for productive uses are among the most critical livelihoods issues. A case on environmental management in a micro-catchment was also chosen because of the critical role environmental resources play in sustaining livelihoods, and because how the environment is managed has a bearing on water resources. The management of urban water services is one of the most important services in cities, and that necessitated the inclusion of a case focusing on that aspect of water resources management. A case on river basin planning was included in the study as it offered an opportunity to have an overarching analysis of how formal institutions managing water resources try to improve livelihoods through planning. Thus the guiding aspect in the selection of cases was the need to highlight water resources management issues which are critical for livelihoods.

1.7.2 Data collection methods

This research made use of the participant observer, focus group discussions and interviews as data collection methods. Participatory research methods, specifically a workshop, were also used to generate data for some aspects of the research.

Participant observer

This research opted for participant observation as the main research technique. The participant observer method leans more towards reflexive science than positivism (Burawoy, 1998), and neither pretends nor actively seeks to engage participants in a neutral way. Instead of seeking to reduce the effects of the researcher on the participant, the method realises that it is impossible to banish such effects completely. Burawoy (1998) argues that one should take the view that context is not noise disguising reality, but is reality itself. In this regard, the research took the standpoint that the context of the research itself would be part of the reality in which the participants live and experience everyday triumphs and struggles. The researcher used the approach of living in the research area for extended periods, observing practices of water users at different sources of water. Of interest to the researcher was how water is obtained and used, and the rules which are observed at the source of water. Participant observer method was also used when the researcher facilitated a workshop at which a ward water plan was made by local water users. The method was also used to study catchment and subcatchment council meetings.

Key informant interviews

Interviews were used to collect data on rules operating at waterpoints within the catchment, household livelihood strategies and the role of formal institutions in water resources management, especially at the sub-catchment, the catchment and at the level of the government Ministry. Key informants were identified with the help of members of the community, and in some cases were chosen on the basis of their position(s) in the community or in organisations they work for. The researcher relied on the unstructured interview technique. In common usage, the term 'unstructured' implies something without form or direction and as such some researchers are reluctant to admit they use an 'unstructured' approach to research for fear of being labelled 'unscientific' because in essence science searches for patterns and is structured. Unstructured interview methods were used because they allow the researcher the flexibility to pursue issues as they arise. Unstructured interviews allowed the researcher to follow up on those issues that came up during the course of the interview. Interviews enabled the researcher to probe water users about aspects of water resources management which are not 'observable' such as the rules which operate at each water point. They were also used to seek clarification on observed practices and other aspects of water resources management which were not clear to the researcher.

Focus group discussions

This research also relied on focus group discussions to gather data. These discussions were restricted to data collection at the local level. The researcher had focus group discussions with irrigators, rainfed farmers, domestic water users and members of water point committees. In the early stages of the research FGDs were used to get insights on the uses of water and the rules that applied at different sources of water. Insights from these discussions were then used to formulate research questions which were used to collect more data from key informants.

Discussions were planned to be held with on average between six and ten participants, but since these took place in public spaces, such as near boreholes, they attracted attention and in some cases as many as fifteen people would be participating. In most cases discussions and interviews were conducted in isiNdebele, which is an indigenous language. Since the researcher is not fluent in isiNdebele, the services of an interpreter were made use of, and proceedings were recorded on a voice recorder.

Participatory research

This study also employed participatory research techniques. Participatory research is when members of the community or a specific group collaborate in the identification of problems, collection of data and analysis of their own situation in order to improve it (Selener, 1997:12). Participatory research was used to collect data on water resources planning (Chapter 7). A workshop was held during which water users identified water resources within the communities, problems they were facing with regards to access to water for domestic and productive uses. Water users then proceeded to suggest solutions to their problems.

1.8 Thesis outline

The cases which were chosen for this study cover an array of aspects of water resources management. Five cases were studied and will be presented in this thesis, and these are: (1) on practices of water users at the local level, with a focus on both domestic and productive water, (2) balancing resource management and livelihoods, with particular reference to watershed management and food availability at the household level, (3) interventions in smallholder agriculture in the management of 'green water' in a dry catchment, (4) struggles for the control of urban water services, and (5) processes of river basin planning within the Mzingwane Catchment. Figure 1.1 shows a schematic overview of how the chapters link.

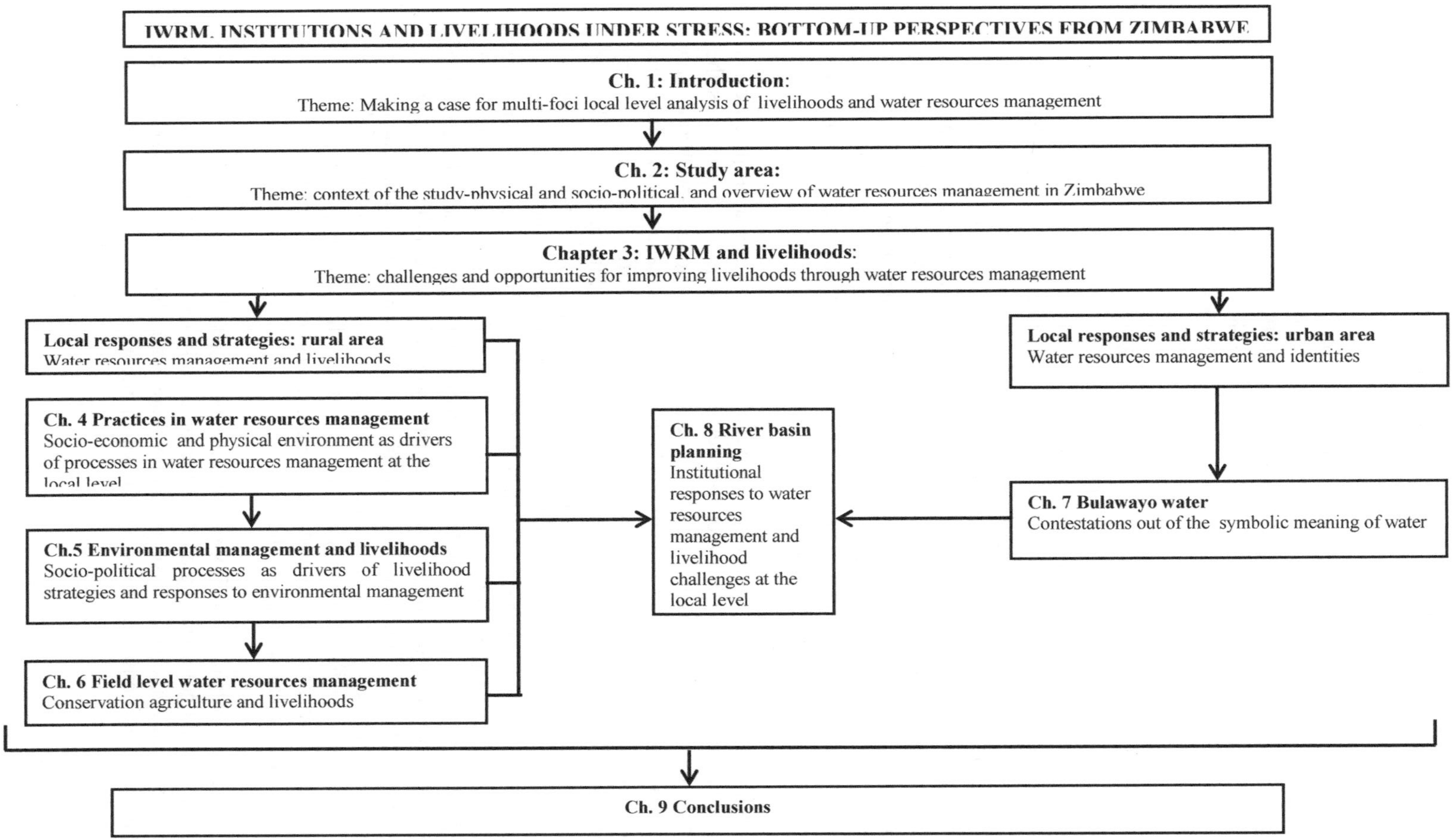
IWRM, INSTITUTIONS AND LIVELIHOODS UNDER STRESS: BOTTOM-UP PERSPECTIVES FROM ZIMBABWE
Ch. 1: Introduction:
Theme: Making a case for multi-foci local level analysis of livelihoods and water resources management
Ch. 2: Study area:
Theme: context of the study-physical and socio-political, and overview of water resources management in Zimbabwe
Chapter 3: IWRM and livelihoods:
Theme: challenges and opportunities for improving livelihoods through water resources management
Local responses and strategies: rural area
Water resources management and livelihoods
Ch. 4 Practices in water resources management
Socio-economic and physical environment as drivers of processes in water resources management at the local level
Ch.5 Environmental management and livelihoods
Socio-political processes as drivers of livelihood strategies and responses to environmental management
Ch. 6 Field level water resources management
Conservation agriculture and livelihoods
Ch. 8 River basin planning
Institutional responses to water resources management and livelihood challenges at the local level
Local responses and strategies: urban area
Water resources management and identities
Ch. 7 Bulawayo water
Contestations out of the symbolic meaning of water
Ch. 9 Conclusions

Figure 1.1 Thesis overview

This thesis is structured as follows: Chapter 1 introduced the research by giving an overview of the study.

In Chapter 2 a description of the study area will be presented. The chapter will describe the agro-ecological characteristics of the study area. The Mzingwane catchment is semi-arid, which limits the availability of surface water for domestic and productive uses. This also limits agricultural productivity. The chapter will also give a brief history of water resources management in the country and an overview of the economic crisis which affected Zimbabwe.

Chapter 3 is based on literature review of Integrated Water Resources Management (IWRM). The chapter takes a closer look at the IWRM framework, giving particular attention to the question whether within the framework it is possible to address livelihood needs. Although IWRM is touted as being able to deliver socio-economic development, there are questions about whether livelihoods are directly targeted within the framework. Conceptual and practical challenges of implementing the framework will be identified in the chapter. The chapter goes further to analyse some of the suggestions which have been put forward to improve the framework. Among these are 'Light IWRM' (Butterworth *et al.*, 2010) which is said to be more feasible to implement in the developing countries. The chapter concludes that the IWRM framework is rendered weak to deliver socio-economic development by that within it water resources development projects are not prioritised. Such projects can potentially improve livelihoods.

Chapter 4, which is the first of the thesis' empirical chapters, is based on practices in water resources management at the water user level. The chapter, which considers the local level to be the ward, makes use of the concepts of practice, interaction and institutional bricolage in its investigation of local water resources management. The focus of the chapter is on how water resources management draws different actors together. The chapter starts by looking at water resources found in Ward 1, Insiza district which is in the Upper Mzingwane subcatchment. The practices of the different water users at different sources of water and water infrastructure will be analysed. The nexus between the manner in which water is managed and the cultural values of the water users will also be established, and be used to show how institutions which manage water are connected to other aspects of the social life. It will be shown that administrative boundaries which are used by development agents in their projects have the effect of dividing communities which are otherwise one. This leads to problems in water resources management. The main argument of the chapter is that water resources management at the local level cannot be understood outside of the context of the everyday lives of the water users. The chapter concludes that practices in water resources management are influenced by a number of factors to enable a generalisation.

Chapter 5 focuses on balancing environmental management and livelihoods. The chapter uses two cases, one on gold panning and another on a gully reclamation project to show how survival strategies among actors at a local level are connected, and are shaped by the physical and socio-economic realities within which the water users are enmeshed. A description of gold panning as a coping strategy is given, showing the reasons why actors engage in the activity, despite the severe damage to the environment. The case of the reclamation of the Gobalidanke gully, which is in the same micro-catchment in which gold panning takes place, will be described. The case analyses the motivation for households to engage in a watershed protection project. The argument put forward in the chapter is that actors, when confronted with an extreme socio-economic and physical environment, dig deep into institutional and natural resources available to them.

Chapter 6 focuses on the link between so-called 'sustainable agriculture', specifically conservation agriculture, and livelihoods. In the chapter it will be shown how conservation agriculture, which among other things aims at enhancing the effective use of rainfall water, draws together smallholder farmers, state and non-state actors. The chapter analyses how and why smallholder farmers align their livelihood needs and strategies with the objectives of non-state actors intervening in agriculture. Non-state actors will also be shown to offer incentives to smallholder farmers which address livelihood needs and therefore encourage farmers to take up conservation agricultural technologies. Although smallholder farmers are not confident about conservation agricultural technologies they do adopt the technologies to safeguard their links with non-state actors and access to benefits associated with these links. The chapter discusses the emerging relations between science and extension, and also between state and non-state actors in smallholder agriculture. The chapter concludes that critical aspects of rural development, such as increased agricultural productivity, cannot be sustained by the use of short-term incentives offered to smallholder farmers.

Chapter 7 analyses the struggle for the control of urban water services. The chapter presents a case study on Bulawayo, which was the site of contestation between central government and the local authority over control of water services. The main research question which the chapter asks is, how does context explain contestations over the management of water resources in an urban area? The chapter also analyses how coalitions emerge over contestations in water resources management. The concept of statecraft (Scott, 1998) will be used as an analytical tool. A brief history of Bulawayo will be presented, to show how the local authority and the residents have approached water resources management over the years. The actual struggle for the city's water services will be analysed , starting with how the government got involved in the city's water services, and how city council resisted the proposed takeover. The chapter concludes that although the government justified the proposed takeover on the grounds of increasing efficiency in service delivery, at stake was its desire to control the city's water services as a way of exerting its control over an urban area. At the time urban areas were perceived as powerbases for opposition political parties. The local authority, which is the Bulawayo City Council, resisted the proposed takeover on the grounds that it did not have a basis in law. However, the opposition to the takeover cannot be separated from that to the local authority and the residents, water services represented a sense of defiance to authority and the will to survive against all odds. The chapter concludes that the symbolic meaning of water was the actual issue at stake in the contestation.

Chapter 8 raises the question, is river basin planning resource planning or development planning, and can it be used to improve livelihoods and local water management? Planning is considered to be central in water resources management. The chapter analyses whether through river basin planning it is possible to align the statutory requirements governing river basin planning to the needs of water users at the local level. The chapter analyses how the socio-economic environment in which planning occurs, and the livelihood needs of water users, force planners to constantly adjust their strategies. A history of river basin planning in Zimbabwe will be given in the chapter, with particular emphasis on the period prior to, and after the water sector reforms in the country. The Mazowe and the Mzingwane catchments, both of which are in Zimbabwe, are compared. Processes which the Mzingwane catchment council used to make its river basin plan will be followed. This is done to find out which actors were involved, or left out of the process of river basin planning, and how different actors were involved. Planning as done by the catchment council is compared to planning as done by water users at the local level. Experiences of the Inkomati Catchment, South Africa, will be used to compare planning processes between South Africa and Zimbabwe. The chapter argues that river basin planning on paper offers an opportunity to bring different actors to the water table. However, this is hindered by the socio-economic realities in which planning occurs. Furthermore, there is a mismatch between the livelihood realities of the water users and the statutory requirements which govern river basin planning.

The final chapter of the thesis summarises the main findings of the research and presents the conclusions reached.

Chapter 2

Study area

2.1 Introduction

This chapter presents the location of the study area. It analyses the agro-ecological setting in which water resources management takes place. The main factors affecting rainfall distribution in the country are discussed and the relationship between drought and socio-economic development in Zimbabwe will be highlighted. The case of the great drought of 1991/92 is used to illustrate this point. The chapter also touches on how the absence of water infrastructure exacerbates the effects of drought in the Mzingwane catchment. A brief analysis of Zimbabwe's political economy, in particular the crisis of the 2000s, is made. The crisis, which was characterised by hyperinflation, had a profound impact on the Zimbabwean society. The socio-history of the Mzingwane catchment is also summarily presented. The Mzingwane catchment falls in the Matabeleland region, which was greatly affected by civil strife which affected Zimbabwe soon after the country's independence. An overview of water resources management in the country is given, focusing on the changes ushered in by the water sector reforms. The reforms were generally based on the principles of IWRM. Among the changes introduced by the reforms was the formation of catchment and subcatchment councils which act as stakeholder participation platforms.

2.2 Location of the study area

This thesis is based on research which was carried out in the Limpopo river basin's Mzingwane catchment, which falls within Zimbabwe. The Limpopo river basin is shared among Botswana, Mozambique, South Africa and Zimbabwe. The basin covers 1.3 % of Africa's total geographical area. The Mzingwane catchment, which is located between 19.8° and 22.4° South and 27.7° and 32.0° East, is one of Zimbabwe's seven catchments[1]. There are four subcatchments in the Mzingwane, which are the Upper Mzingwane, the Lower Mzingwane, Shashe and Mwenezi. The catchment covers approximately 63, 000 km^2. When using administrative boundaries, the Mzingwane catchment generally falls within Zimbabwe's Matabeleland South Province. Matabeleland South is in the south-western part of the country. The provincial capital is Gwanda, which is a mining town about 600 km south-west of Zimbabwe's capital city, Harare. The province has seven districts which are Bulilima, Mangwe, Matobo, Insiza, Gwanda, Beitbridge and Umzingwane. The bulk of the study, covered in Chapter 4, 5 and 6, was carried out in Upper Mzingwane's Zhulube micro-catchment, which administratively falls in Ward 1 of the Insiza District. Figure 2.1 shows the location of the Insiza District in Zimbabwe.

---

[1] The Manyame, Mazowe, Gwayi, Sanyati, Runde and Save catchments are the other six.

Chapter 7 takes a catchment-wide approach, and Figure 2.2 shows a map of the Mzingwane catchment and its subcatchments.

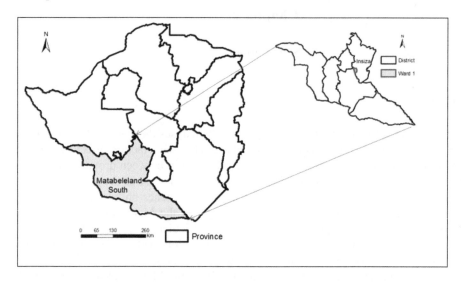

Figure 2.1 Location of Matabeleland South Province, the Insiza District and Ward 1

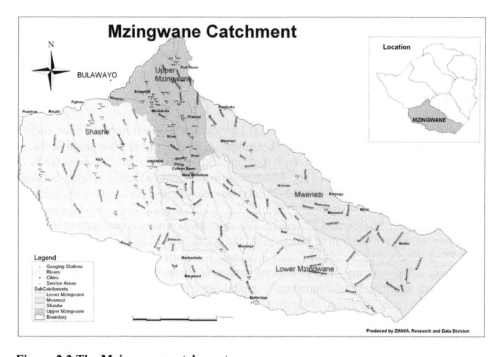

Figure 2.2 The Mzingwane catchment

2.3 The agro-ecological setting

Zimbabwe's rainfall is strongly related to the seasonal movements of the Inter-Tropical Convergence Zone (ITCZ). The country receives summer rains, occurring from the end of October to early March. This is when the ITCZ is above Zimbabwe (De Groen, 2002). Winter, which occurs between May and August, is dry. The northern parts of the country tend to receive more rains than the southern parts, although local relief plays a role in influencing rainfall type and pattern. This rainfall gradient explains why the Mzingwane catchment, which is located in the southern part of the country, is one of the driest catchments in the country. The north-south rainfall gradient is also consistent within the Mzingwane catchment itself with the northern-most subcatchment, the Upper Mzingwane, being wetter than the southern-most subcatchment, the Lower Mzingwane. To illustrate this, annual rainfall at Esigodini (Upper Mzingwane) ranges from 200-1200 mm/a over the last 70 years, while at Beitbridge (Lower Mzingwane) it ranges from 50-500 mm/a (Moyce *et al.*, 2006). In the Zhulube catchment, which is in the Upper Mzingwane subcatchment, mean annual rainfall is about 550mm/a (Love *et al.*, 2010). Although this rainfall pattern to an extent explains some of the challenges relating to agricultural production, the region also suffers from periodic extreme weather events which affect Zimbabwe in particular and the southern Africa region in general. One such event was the 1992 drought, which is described in Box 2.1.

Box 2.1 The impact of the drought of 1992

In 1992 Zimbabwe experienced its worst drought in living memory (Maphosa, 1994). The drought was not restricted to Zimbabwe but it also affected a number of countries in southern Africa. Within the region received rainfall was below normal by as much as 80%, and this had a huge impact on agricultural productivity (Zinyowera and Unganai, 1993). Zimbabwe was transformed from a food surplus position to being a net food importer (Maphosa, 1994). The Grain Marketing Board (GMB), which is a grain buying parastatal, received only 13 000 tonnes of maize during the year, which was just about enough to feed the country for two days (Maphosa, 2004). More than a million cattle, about a quarter of the national herd, died (Eldridge, 2002). The most devastating impact of the drought was at the household level. Most households only harvested food enough for just about 2-3 months. Apart from impacting on household food security, the drought also affected the Gross Domestic Product (GDP), this being a result of the strong relationship between agriculture and the economy.

Most of the rivers in the Mzingwane catchment are ephemeral, flowing only for a few months or even less during the rainy season. This severely limits the importance of rivers and streams as sources of water for both domestic and productive uses. However, dams have been constructed across the major rivers in the catchment such as the Umzingwane, Mwenezi and the Insiza Rivers. These dams are mainly used for irrigation and to supply urban settlements with water. Through inter-basin transfers, for example, the city of Bulawayo which is in the neighbouring Gwayi catchment gets water from the Mzingwane catchment. However, a sad reality in the catchment is that a sizeable proportion of smallholder farmers are unable to utilise water stored in dams simply because there is no infrastructure to convey water from the dams to the farmers' fields. Box 2.2 describes a condition referred to as a wet drought, which is common in smallholder farming areas in the Mzingwane catchment. Although surface water is scarce in the catchment, groundwater is an important resource in the catchment. Throughout the catchment boreholes are an important source of water for domestic uses and at the peak of the dry season can be the only source of water available to households. In some parts of the catchment, such as the Lower Mzingwane subcatchment, households make use of water trapped in riverbeds for both domestic and productive uses.

Box 2.2 Wet drought in the Mwenezi subcatchment area

One of the most critical factors limiting human socio-economic development in the Mzingwane catchment is the shortage of water. As has been mentioned earlier, this is due to prevailing climate in the catchment. However, human factors have also contributed towards limited socio-economic development in the catchment. The situation in the Mzingwane is an example of what is commonly referred to as a 'wet drought.' A wet drought can be considered as being caused by a combination of socio-economically and natural factors. It occurs when, because of insufficiently developed water infrastructure, water users are unable to make use of available water resources and therefore are unable to deal with natural conditions such as low rainfall (see, for example, Mehta, 2001). In the Mzingwane catchment this is illustrated very well in the Mwenezana subcatchment. Within the subcatchment is a large dam, the Manyuchi Dam, which is across the Mwenezi River. The agreement between the state and the private company which developed the dam is that 13.33% of the dam's water yield should be allocated to surrounding communities. However, to date there has been very little infrastructural development to enable communities within the vicinity of the dam to make use of the water set aside for them. Some of the communities which could make use of the Manyuchi Dam include Maranda, Neshuro, Matibi 1 and 2, Furidzi, Mberengwa, Makuwerere, Mataga, and Gwatemba. This absence of infrastructure to enable the communities to tap the water of the dam results in the communities bearing the brunt of low rainfall totals which could be mitigated by utilising water from the dam.
Source: Love *et al.* 2009.

In Ward 1 soil types vary from clays and loams in the north to sandy soils in the south with stony high ground (Tunhuma *et al.*, 2007). Generally the soils are infertile and shallow, which contributes towards low crop productivity in the catchment. Farmers try to improve soil fertility by applying livestock manure. Use of crop residues as manure is done on a very small scale because low crop yields mean little crop residue is available. Furthermore, smallholder farmers prefer to use crop residues as livestock feed. Use of synthetic fertilizers is also very low because factors related to availability, pricing, and access to the market. One can argue that the challenges of crop productivity in the catchment are self-reinforcing and the smallholder farmers are caught in a poverty-trap.

The most common type of vegetation in the semi-arid areas of southern Zimbabwe is sweet veld, with comparatively high nutritional value of browse and annual grass species. However, significant proportions of the rangelands are degraded, resulting in low biomass and thus limited feed resources of poor quality particularly during the dry season.

Having looked at the agro-ecological setting, which has a strong bearing on the availability of water as a physical resource, the following section analyses Zimbabwe's political-economic environment, which has an important bearing on water resources management.

2.4 Zimbabwe's political economy

In the recent past Zimbabwe experienced one of the most turbulent times in its post-colonial history. The Zimbabwean crisis was characterised by a serious economic decline and social and political turmoil. A characteristic of the period soon after Independence in 1980 was positive economic growth (Richardson, 2005). Exceptional years were 1984 and 1992 during which economic growth was affected by serious droughts (Makochekanwa and Kwaramba, 2009). However, in the 2000s the country's economy was characterised by negative economic growth with inflation being a major problem. By January 2009, for example, annual inflation was estimated to be about 5 billion per cent, while month on month inflation was about 230 million per cent (*ibid*)[2]. An indicator of how inflation eroded the Zimbabwean currency is that by 2009 the highest Zimbabwean denomination was Zw$100 trillion, which on the black-market fetched less than US$10.

The crisis affected basic public services such as education, health, electricity and water supply because the government simply did not have the financial resources to sustain such services. Factors, most of them linked to the economic crisis, such as an ageing infrastructure, shortage of water treatment chemicals and erratic power supply, combined resulting in a serious cholera outbreak from August 2008-2009. The outbreak had its epicentre in high density suburbs of Harare and resulted in tens of thousands of cases and about 4000 deaths (World Health Organisation, 2011). Since the cholera outbreak was mostly concentrated in the urban areas it had the effect of focusing attention on urban water services, to an extent at the expense of the same sector in the rural areas. The government attempted to deal with the problem of water supply and quality of water in urban areas by directing that all urban areas hand over their water services to the Zimbabwe National Water Authority (ZINWA). The collapse of the manufacturing industry resulted in massive unemployment, and by 2003 it was estimated that 80% of the population was living in poverty. As the crisis worsened population movements out of the country intensified. Estimates put the number of Zimbabweans in South Africa alone at more than a million (Kramarenko, 2010). Within the Mzingwane catchment these developments affected smallholder productivity by making agricultural inputs scarce or pricing them beyond the means of the smallholder farmers. This worsened food shortages.

Politically the government faced internal and external pressure. Results of the 2008 elections were contested by the opposition political parties. The electoral process was generally considered to have been flawed. Violence which broke out in some parts of the country was one of the major reasons why the elections were considered not to have been free and fair. Although some regional organisations refused to condemn the electoral process, most western countries and regional groups such as the European Union refused

---

[2] There is no agreement as to what the rate of inflation was at its worst, between 2008 and 2009, because it had become almost incalculable. However, the consensus is that, outside of war zones, during that period Zimbabwe had the worst inflation in the world.

to recognise the outcome of the elections. Sanctions were imposed against the country which affected the flow of aid and Direct Foreign Investment (DFI).[3]

There is no consensus as to what caused the Zimbabwean crisis. The government blamed the crisis on 'illegal' sanctions imposed on the country and a series of droughts which allegedly affected agricultural productivity. However, one of the most popular theories is that the Fast Track Land Redistribution Programme which the government embarked on in 2000 (Zikhali, 2010) started the chain of events which culminated in the crisis. The land reform programme created insecurity in property rights which made the country an unsafe destination for capital (Richardson, 2005). Richardson furthermore refutes the argument that droughts, such as the one of 2001-2002, started the economic crisis. He argues that in the past Zimbabwe had not experienced such economic decline even when it suffered droughts worse than the 2001-2002 one. The argument by Richardson is opposed by Andersson (2007) who argues that food insecurity in the aftermath of the land redistribution programme is more complex than the issue of property rights. Economic misrule and corruption are also some of the theories which have been given to explain the economic turmoil (Makochekanwa and Kwaramba, 2009).

The above section analysed Zimbabwe's political economy, with emphasis on the economic crisis which affected the country in the 2000s. This provides the background needed to understand water resources management in the country. However, it is important to note that the Mzingwane catchment, as part of the Matabeleland region, has a special historical context which needs to be discussed for a better understanding of the relation between central government and the region. In the coming section the civil strife which affected Matabeleland in the 1980s and the socio-economy of the region is discussed.

2.5 The socio-political economy of Mzingwane catchment

The Ndebele are the majority ethnic group in the Mzingwane catchment, which in part is explained by historical factors. The Ndebele were part of the Tshaka's Zulu kingdom and moved into what is now Zimbabwe during the time of the Mfecane in the early part of the 19th century. The Mfecane was a time of turmoil in the Zulu kingdom. The Ndebele made their capital near what is modern-day Bulawayo, and from there they spread into the southern parts of Zimbabwe. However, apart from the Ndebele there are also Shona speaking people and people who originated from neighbouring countries who reside in the region. Some of these came as migrant workers in the mines within the catchment. Table 2.1 presents the demographic information of the study area.

---

[3] The United States and the European Union claimed to have imposed what they termed as 'targeted sanctions' against key members of the government and ZANU P.F., but the government claimed that the sanctions were targeted against the country and its innocent citizens.

Table 2.1 Demographic indicators for the study area

	Matabeleland South	Insiza District	Ward 1
Male	328 000	49 000	1900
Female	357 000	51 000	2000
Total	685 000	100 000	3900

Source: Zimbabwe National Statistics Agency (2012)

Civil strife in the Matabeleland region

One cannot describe the history of Matabeleland in Zimbabwe's independence era without reference to the civil war which occurred in the region in 1980s. Matabeleland region, together with the Midlands province, experienced a civil war, which has come to be known as Gukurahundi in the period soon after Zimbabwe's independence. According to the Catholic Commission for Justice and Peace in Zimbabwe (1997) the disturbances, which started in February 1981 with a violent outbreak at Entumbane in Bulawayo, have been traced to the suspicions between the newly elected Zimbabwe African National Union (ZANU) government and the Zimbabwe African People's Union (ZAPU). A split in the liberation movement resulted in the formation of ZANU. This resulted in the Rhodesian government fighting two guerrilla movements, ZAPU Patriotic Front and ZANU Patriotic Front. In the first national elections ZANU won the majority vote, but became suspicious that ZAPU wanted to unseat the government through unconstitutional means. These suspicions were heightened by the discovery of arms caches in the Matabeleland region. Subsequent banditry activities in the region, which were attributed to ZAPU supporters, resulted in a military response from the government. Estimates put the number of deaths due to the disturbances at as high as 20 000, while other impacts of the war included physical and psychological problems in the people of the region. As much as 200 000ha of commercial agricultural land was abandoned as the war made farming activities difficult. The civil war remains a dark spot on the country's history, particularly because the government has never formally apologised for the atrocities attributed to the army. To date the people of the region see under-development in the region as part of a continuation of the marginalisation of the Ndebele. Apart from the massive loss of life and economic impact of Gukurahundi, the civil war also affected the construction of ward boundaries in Matabeleland, which is described in Box 2.3.

Box 2.3 Impact of Gukurahundi on the construction of boundaries

The Prime Minister's Directive of 1984 created Village Development Committees (VIDCOs) and Ward Development Committees (WADCOs). These were meant to improve administrative and governance efficiency (Makumbe, 1996). However, Alexander (2006) states that these never really gained popularity in Insiza because of the manner in which they were formed, and the perceived purposes for which they were formed. Wards were formed during the time of Gukurahundi in Matabeleland region, which has been described above. Because of the civil war, security issues were high on the political agenda. As a result the Local Government Promotional Officer (LGPO) tasked to draw VIDCO and ward boundaries mapped them from his office due to the security threats (*ibid*). Boundary wards which were created therefore did not take into account patterns of resource use or social ties among the people. At the time the government was desperate to contain dissidents and to try and create a one-party state, so communities perceived VIDCOs as being part of the surveillance strategy of the central government. Thus VIDCOs were boycotted as people were reluctant to serve on them or support them.

Economic activities in the Mzingwane catchment
With the exception of Beitbridge, which is a frontier town, most of the Mzingwane catchment is based on primary industrial activities of mining and agriculture. Urban settlements such as Gwanda, Coleen Bawn, and West Nicholson, for example, are historically linked to these industrial activities. Gold is the most important mineral in the catchment, and is mined at a commercial scale and also at the level of individuals as a household coping strategy.

Commercial farming, especially the growing of citrus fruits, is an important economic activity in parts of Beitbridge District, which is in the Lower Mzingwane subcatchment. Citrus fruits are exported to South Africa. The area around West Nicholson and Coleen Bawn is renowned for commercial cattle ranching and has a well-developed meat processing industry. Vast tracts of land within the catchment are also used as wildlife sanctuaries. This attracts tourists to the region. However, one has to add that after the country's land reform programme commercial farming has declined and there have been threats to take over some wildlife sanctuaries. As far as livestock production is concerned, estimates put the decline of the national herd due to the land reform programme at 75% between 1996 and 2004 (Homann *et al.*, 2007).

In the communal areas agriculture is an important livelihood activity, but to say that it is an important economic activity would be rather difficult. This is so because a combination of agro-ecological and socio-economic factors militates against meaningful agricultural production. Households struggle to produce sufficient food to meet their own needs and have to rely on food aid most of the times.

Livestock production is one of the major activities in the catchment. A study by Homann *et al.* (2007) found that goats were increasingly being used to supplement household income and to enhance food security. The study also found that poor access to animal health support, dry season feed shortages, and inadequate housing were the major factors affecting livestock mortality. Goat ownership is very high among smallholder farmers. At least 40% of the households do not own cattle, so goats complement their livestock resources. Unfortunately it appears that farmers are not realizing the potential of selling goats. Homann *et al.* found that there were no formal market facilities for selling goats and as a result smallholder farmers relied on farm gate sales. Without a well-established marketing system the smallholder farmers are not realizing the full potential of their livestock.

Although the majority of smallholder farmers in the catchment practice rainfed farming, there are a number of irrigation schemes. There are small schemes with less than 20 plotholders, each having less than a hectare of land. The Wabayi Irrigation Scheme near Gwanda is one such example. There are also large schemes with close to a thousand plotholders such as the Silalabuhwa Irrigation Scheme with an irrigated area of about 450 hectares (Watermark Consultancy, 2010). Most of the irrigation schemes in the catchment were funded either by the state or by donor-agencies. In some cases plotholders receive inputs from NGOs, and during the government-sponsored *Operation*

Maguta the government also provided inputs. Such schemes in a way contribute towards food security in the catchment to the irrigators, and indirectly to the community who buy produce from the scheme. As with livestock farming, plotholders also face the problem of how to access markets outside the immediate communities.

2.6 An overview of water resources management in Zimbabwe

The history of water resources management in southern Africa is closely related to political economic factors dating back to the period of colonisation (Swatuk, 2008). This is also true for Zimbabwe which was a British colony. From the time of colonisation in 1890 until 1998 when the post-independent government reformed the water law, water resources management in the country was meant to serve the interests of the European settlers. The interests of colonisers were to develop a settler economy through resource extraction (Manzungu and Machiridza, 2005; Swatuk, 2008). As a result, between 1890 and 1927 the focus of the newly established colony was mining, thus water was committed to that sector (Manzungu and Machiridza, 2005). However, when the settlers failed to find as much minerals as they had anticipated their attention turned from mining to agriculture. That shift ushered in the era of land dispossession from the Africans. It is important to note that Africans were moved to marginal areas where agricultural potential was limited by low rainfall and poor soils. Whites occupied areas of high agricultural potential. As the economic base of the country changed, the water law was changed accordingly to meet the needs of the agricultural sector. In the Water Act (1927) water rights were attached to land following the adoption of the riparian doctrine (Bolding, 2004). Water rights were issued based on the Priority Date System (PDS). The same Act also prohibited cultivation within 30m of streambanks (*ibid*). Colonial water law made it difficult for Africans to access water for productive uses. This was so because water rights were attached to land ownership, yet by law indigenous people did not have title to land. The PDS also meant that Africans could only be allocated water after the needs of whites, who had priority by virtue of being earlier applicants, had been met. Racial policies of the settler-led government also facilitated the development of water resources in areas under white occupation while water infrastructure in the areas set aside for Africans was neglected (Campbell, 2003). It is in these racially biased policies of the colonial regime that the roots of African poverty, particularly in the rural areas, are found.

Although the colonial water law was carried on into the independent era it was clear that the policies of the colonial state could not be sustained in modern Zimbabwe. Consequently in 1993 moves to reform the water sector started. The reforms, which culminated in the enactment of the 1998 Water Act, ushered in the era of water resources management within the framework of Integrated Water Resources Management (IWRM). However, water sector reforms were not isolated but were linked to macro-economic changes which the country was implementing at the time (Chikozho, 2008; Mtisi, 2008). At the time the country was in the process of implementing economic structural adjustment programmes, whose major aims were to reduce government expenditure. Decentralisation was also encouraged by the International Financial Institutions (IFIs) which were the instigators of the adjustment programme. It is therefore not a coincident that water sector reforms resulted in a decentralised institutional set-up. Major changes which came about as a result of the water sector reforms include the formation of the Zimbabwe National Water Authority (ZINWA), catchment councils (CCs) and subcatchment councils (SCCs). Management of water resources was also shifted from administrative boundaries to hydrological ones.

The national water authority, which was created by the ZINWA Act (1998), was formed to be a parastatal tasked with providing a framework for the development, management, utilisation and conservation of water resources in the country (Gumbo, 2006, Mtisi, 2008). It replaced the Regional Water Authority (*ibid*) and took over the functions associated with the provision of raw water (bulk water) which used to be the responsibility of the Department of Water Development (DWD). Sale of raw water earns the parastatal revenue. Other functions of the national water authority include:

- Institutional coordination in the water sector
- Assisting catchment councils to discharge their duties
- Maintaining dams previously owned by the government
- Undertake research and develop a data base on hydrological issues
- Promote mechanisms for the cooperative management of international water resources
- Advice the Minister responsible for water on the formulation of national policies and standards on dam safety, water pricing, water resources planning and management and development

Catchment councils and subcatchment councils were created by the Water Act (1998), and these were formed to be stakeholder institutions responsible for the day-to-day management of water resources. Within the organisational hierarchy in the water sector, subcatchment councils are the lowest legal entity. They are comprised of stakeholder group representatives. Statutory Instrument 47 of 2000 identifies some of the stakeholder groups which can be included in subcatchment councils as including Rural District Councils (RDCs), urban authorities, communal farmers, resettled farmers, small and large scale commercial farmers, and small and large scale miners. Among the functions of SCCs are to:

- Regulate and supervise the exercise of water permits
- Monitor water flows and water use
- Assist in data collection and participate in planning for water resources
- Collect levies from water users

Catchment councils are in turn composed of representatives from subcatchment councils, in general these being the chairperson and the vice chairperson of each subcatchment council. The main functions of catchment councils include to recommend the apportionment of water, specify major uses of water, draw up an inventory of water resources and to supervise the functioning of subcatchment councils below them. Chapter 7 which looks at catchment planning and the search for institutional legitimacy will make an analysis of the functioning of the national water authority and catchment and subcatchment councils. Notably, in the post-water reform framework the Department of

Water Development's (DWD) main task is to formulate national policies and standards for management and development of water (Gumbo, 2006).

The period following the water sector reforms in the country was characterised by a drastic change in the fortunes of Zimbabwe. The country went from being able to produce enough food to meet its needs to having to import food, from a thriving economy to a ruined economy (Richardson, 2007). By 2001, for example, direct foreign investment which had contributed to economic growth, had dwindled to almost zero, and by 2003 the economy was shrinking faster than any other in the world (*ibid*). Futhermore, the Zimbabwean dollar had lost more than 99% of its value. This had a profound impact not only on the populace, but also on water resources management.

2.7 Summary

This chapter gave an overview of the study area as a way of providing the socio-economic and ecological context in which water resources management takes place in the Mzingwane catchment in particular and in Zimbabwe in general. Arguably, the country's socio-economic development has always been predicated on water resources management given the importance of water to agriculture, and that of agriculture to the economy. In the pre-independence era racial policies focused on white possession of water and land as a way of ensuring high agricultural productivity. Blacks were displaced to areas of low rainfall and low agricultural productivity and were denied a seat at the water-table through policies which tied access to water to land ownership. The pattern of poverty distribution in the country today has its historical origins in these factors. It can be argued that in the pre-independent era the relations between the government and the indigenous smallholder farmers as a special group of water users were characterised by racial divisions between whites and blacks. However, while the racially motivated policies which militated against smallholder farmer productivity have been dismantled, it is notable that a new set of factors has come into play and which is now affecting productivity. The economic collapse of the country, for example, weakened the government to such an extent that it could no longer provide basic services necessary for the running of the country. The government could also not meaningfully intervene in smallholder farming and water resources management in particular. In the Mzingwane catchment such challenges which occur at the national level are exacerbated at the household level by natural factors such as low rainfall totals and an erratic distribution of rainfall within the season. These factors, combined with low natural levels of soil fertility, lead to low agricultural productivity resulting in food insecurity at the household level. This background suggests that interventions in water resources management have to take into account multiple factors, some of which may lie outside what has traditionally been considered to be the domain of water resources management.

Chapter 3

Integrated Water Resources Management and livelihoods: do they meet?

3.1. Introduction

Southern Africa is characterised by socio-economic under-development which is linked to natural, socio-economic and political factors (Swatuk, 2008). About 70% of the region's population subsist on less than US$2/day while 40% live on less than US$1/day (*ibid*). Millions of people in the region have limited access to water, which is critical for human welfare. A sad reality is that inequities in access to water are linked to the manner in which powerful political and economic interests have managed water resources (Swatuk, 2008). Pre-independent governments in southern Africa developed water infrastructure and allocated the resource in such a way that areas occupied by whites were better served than those occupied by blacks. Such socio-political processes, combined with natural factors of low average rainfall, to a large extent account for the limited access to water by the poor. Limited access to water contributes towards, among other things, low crop productivity which in turn results in food insecurity. Arguably, one of the major challenges which confronted independent governments in southern Africa was that of improving water resources management for the purpose of improving livelihoods.

Over the years initiatives have been taken to try and improve water resources management, and generally these have directly or indirectly targeted improving livelihoods. Between the 1950s and the 1970s, for example, during the era of the hydraulic mission, the focus of water resources management was large dam construction (Allan, 2003). The construction of large water infrastructure was mainly to meet the needs of industry, agriculture and the energy sector. Livelihoods would benefit through increased crop productivity in irrigation schemes, or through employment created by an expanding manufacturing industry, for example. Hydraulic projects which increased electricity generation capacity would indirectly benefit water users through an improved economy. However, the sad reality then was, and continues to be that the focus of water resources management is on stored water at the expense of, for example, rainfed farming (Van der Zaag, 2005). This suggests that whereas other sectors and stakeholders benefit directly from water resources management projects, rainfed farmers at most are indirect beneficiaries of initiatives in water resources management. This jeopardises the livelihoods of rainfed farmers who constitute the majority of rural water users in Zimbabwe just as they do in most countries in southern Africa.

However, it is notable that in the recent past a water resources management framework which claims to facilitate socio-economic development and improve the conditions of the poor in a more direct manner has emerged. The framework is known as Integrated Water Resources Management (IWRM) and is being promoted by international actors such as the Global Water Partnership (GWP). IWRM is defined as a process through which water, land and related resources are managed in an integrated manner for maximising socio-economic and ecological benefits (GWP, 2000)[4]. Implementing IWRM is said to facilitate the attainment of 'Three E's' which are (social) equity, ecological integrity and economic efficiency (*ibid*). Such claims have raised hopes that through water resources management it can be possible to improve livelihoods. This explains why the majority of countries in southern Africa have adopted the framework, revising their water laws and policies in line with Principles of IWRM and setting up organisations suggested in the IWRM framework. Although southern Africa in general, and Zimbabwe in particular, has been quick to implement IWRM, a major question in need of answers is, can IWRM in its current format contribute towards improving livelihoods? This chapter focuses on that question.

The chapter is structured as follows: firstly the analytical framework which is adopted in the chapter is presented. The chapter makes use of the livelihoods perspective which has been defined by Chambers (2004) and is commonly used by development agents. This perspective is adopted because improving livelihoods is critical in southern Africa given the levels of poverty in the region. Water resources management can potentially contribute towards improving livelihoods. Secondly the chapter focuses on some of the conceptual and operational aspects of IWRM. Integration is one aspect of IWRM which is analysed because of the conceptual and operational challenges it poses. Thirdly attention turns to suggestions which have been put forward for improving IWRM. Some of the challenges address perceived conceptual shortcomings of IWRM, such as what should be integrated within the framework. Such suggestions include that IWRM should be informed by Integrated Natural Resources Management (INRM) and be strategic and adaptive. The section will also examine suggestions which try to address the practical challenges of implementing IWRM in such a way that livelihoods benefit. In particular the multiple use systems (MUS) and 'Light IWRM' are approaches which have been put forward to try and improve IWRM. Finally a discussion and conclusions are presented. The chapter makes the observation that the sad reality is that debates on how to improve water resources management seem to eclipse the need for delivering tangible benefits to the poor. There is need for more practical-oriented solutions to the challenges which the poor face. Since the causes of poverty, which is a major challenge in southern Africa, are multiple, it is ambitious to expect a single sector (water) to solve these problems alone. However, a concerted effort towards understanding the livelihoods-water nexus at the local level can be the first step towards solving such challenges.

3.2 The rise of the livelihoods approach

One of the major challenges confronting southern Africa is that of precarious livelihoods. This challenge needs urgent attention, and one school of thought argues that if resource management (water in particular) takes a livelihoods-centered approach some progress towards improving livelihoods can be made (Merrey *et al.*, 2005). The basis of this

---

[4] Snellen and Schrevel (2004) consider the GWP definition to be the first authoritative definition of IWRM. However, there has been a steady proliferation of definitions of IWRM offered by different actors.

argument is that, although there are claims that implementing IWRM can contribute towards socio-economic development, there is no explicit reference to livelihoods within the 'official' definition of IWRM as given by the GWP. The absence of such reference has raised the following questions: can livelihoods be improved within the IWRM framework? Is improving livelihoods a goal of water resources management? It has also been argued that the primary purpose of water resources management should be managing people's livelihood activities so that the water cycle and the environment in general are not disrupted by these activities (Jonker, 2007). If these perceptions are applied to water resources management it can be argued that IWRM must be conceptualised and implemented in such a way that livelihoods are at the heart of the framework.

The livelihoods framework, which has been adopted by international development agencies such as the United Kingdom's Department for International Development (DFID) rose out of the disappointment with purely economic analyses of socio-economic development (Huckle and Martin, 2001). Such analyses focused on productive resources, such as finance and labour which could be quantified and valued in monetary terms. Although such econometric approaches had, and still have an important role in understanding socio-economic development, it was realized that they did not capture the full essence of livelihoods. This was particularly true for rural livelihoods where, for example, institutional aspects of resource access and use, are pivotal in shaping the quality of livelihoods. Such 'unquantifiable' factors play an important role in determining the opportunities available to households, and in turn their productivity. These factors led to the search for a framework which could go beyond analyses of quantifiable resources but could capture rural livelihoods in their entirety.

Within the livelihoods framework a sustainable livelihood comprises people, their capabilities and their means of living, including food, income and assets (Chambers and Conway, 1991). Tangible assets can be resources, while intangible assets can be claims and access. Access refers to the real opportunity actors have to use or benefit from a certain resource. A livelihood is considered to be sustainable if it can cope with and recover from stresses and shocks and provide for future generations. The importance of such a definition of livelihoods is that it identifies key areas which affect human welfare, which include the opportunities people have, and critical aspects of human welfare such as such as food security and income. Taking this view of livelihoods to water resources management suggests that it is not enough to analyse the economic aspects of water resources management, such as the cost of supplying water, but there is need to understand the institutional mechanisms through which water is accessed. This might necessitate, for example, an analysis of the power dynamics of how rural water users gain access to sources of water, and how that enhances or hinders the improvement of livelihoods.

The livelihoods framework has been adopted by international development agencies which directly or indirectly, fund developmental programmes in developing countries. Notably, in Zimbabwe the DFID has been responsible for funding relief efforts and programmes related to conservation agriculture. Organisations funded by such donor agencies frequently evaluate the success or failure of their projects using the livelihoods framework, for example, by looking at how projects have changed access to resources, or have changed household income.

For a region such as southern Africa which has a high incidence of food insecurity and low income by global standards, a livelihoods-centered approach can potentially facilitate the setting of a socio-economic development agenda which can touch on the various aspects of rural lives. However, it is important to recall that a livelihood comprises a number of facets. With this in mind it is critical to ask, is it possible for IWRM to 'single-handedly' deliver socio-economic development, especially taking into account the multiplicity of facets that make up livelihoods?

3.3 Can IWRM deliver socio-economic development?

As mentioned earlier, proponents of IWRM suggest that implementing the framework can facilitate socio-economic development (GWP, 2000). However, these claims have been questioned on a number of fronts, such as the operational challenges of implementing IWRM, and the conceptual inconsistencies of the framework. These are discussed in this section looking at integration in particular.

3.3.1 Operational challenges in IWRM

Integration is one of the most contested aspects of IWRM, and it is also one of the most critical elements of water resources management if livelihoods are to be improved. Within the IWRM framework a lot of emphasis is placed on integration (Garcia, 2008). The GWP-TAC (2000) states that integration should be considered under two categories, which are the natural and the human system. The natural system is important for resource availability and the quality of the resource while the human system determines resource use, the generation of wastes, and above all, is responsible for setting the development agenda. Table 3.1 shows possible areas of integration as suggested by different authors.

Table 3.1 Possible areas of integration in IWRM

GWP (2000)	Mollinga (2000)	Keen (2003)	Grigg (2008)
1. Land and water management	1. Different uses of water	1. Uses and objectives for water management	1. Policy sectors
2. 'Blue' and 'green' water	2. Analytical and policy perspectives	2. Human and natural systems interaction	2. Water sectors
3. Upstream and downstream water related interests	3. Different institutions	3. Sectoral policies	3. Government units
4. Cross-sectoral integration in national policy development	4. Geographical integration		4. Organisational levels
5. All stakeholders in the planning and decision making process	5. Water resources development and management and rural transformation and poverty alleviation		5. Functions of management
			6. Geographic units
			7. Phases of management
			8. Disciplines and professions

Given the nature and range of issues which have been suggested for integration, critics of IWRM argue that it is operationally impossible for the framework to be implemented (Biswas, 2004). Critics of IWRM have identified as many as 35 sets of issues which can potentially be integrated. These range from water quality and quantity, water and land issues, and present and future technologies (*ibid*). Attempting to integrate all these variables is not only time consuming because of the wide range of issues they cover, but can potentially result in an unwieldy institutional structure, if at all the integration can be done (*ibid*). Although integration stands out as one of the cornerstones of IWRM, it is not clear how use, development and management of land and water resources can be integrated (Biswas, 2004). Furthermore, integration is likely to result in water managers having to deal with issues that are beyond their expertise (*ibid*). This might not auger well for (water) resource management. Attempting to achieve integration as suggested within the IWRM framework can also potentially paralyse action as the institutional coordination and changes required to operationalise the integration are not easy and are time consuming to put in place. This time-delay can potentially jeopardise livelihoods. However, a suggested solution to this problem is that at the strategic level the broadest range of variables and their inter-relationships can be considered, but at the operational level a more focused approach is taken (Mitchell, 2005; Lankford *et al.*, 2007). This would potentially make integration possible and IWRM feasible.

Conceptual inconsistencies in IWRM

However, it is not only at the operational level that IWRM has been criticised, the conceptual soundness of the framework has also come under scrutiny. Sceptics argue that IWRM is in effect a 'woolly' and 'fuzzy' concept which ultimately is not very useful for water resources management (Molle, 2008). The framework almost has an emotional attraction which different actors are willing to embrace. Critics argue that, for example, the concept of 'integration' is appealing to different actors as it gives the impression of moving from a state of fragmentation and chaos to a state of better water resources management (Molle, 2008). Given the real and perceived conflicts in the water sector, another weakness of integration has been given as that it is a consensual concept (*ibid*). Integration gives the impression that by bringing together different actors it can be possible to avoid inter-sectoral conflicts. Thus integration presents an idealistic vision in which different actors meet to discuss water allocations and share the resource in harmony. Unfortunately the idealistic nature of integration seems to ignore real world challenges of how to resolve the power differentials among actors. In the real world water resources management, by its very nature, brings together actors from different backgrounds, with different agendas and priorities, which makes it extremely difficult for the ideals of integration to be achieved. Thus, although integration is an attractive option to (perceived) chaos and dis-harmony in water resources management, a potentially useful concept is one which offers a way of dealing with the realities of conflict and competition and power differentials among water users. In Zimbabwe it has been shown that during consultation meetings commercial farmers are protective of their interests and smallholder farmers do not wield enough power to tilt negotiations in their favour (Sithole, 2001). In the end the needs of smallholder farmers are not considered and that affects participation. This can potentially result in the status quo being maintained in water resources management and therefore bring no livelihood benefits to the rural poor.

Interestingly, integration has also been criticised from the standpoint that, judging by the manner in which it is being promoted in the IWRM framework, it is not all embracing. Critics point out that integration within IWRM does not take a truly holistic view of natural resources (Merrey *et al.*, 2005). They argue that land resources are only included superficially within the framework, and there is need for better integration of land and water resources management (*ibid*). In addition, other natural resources which are critical for rural livelihoods, such as forest resources and biodiversity, appear not to be included in the framework. Furthermore, issues of access to land and markets, and provision of infrastructure such as roads and telephone networks should also be integrated into water resources management (Merrey *et al.*, 2005). These elements do indeed make the fabric of rural socio-economy and attending to them can potentially improve livelihoods.

For countries such as Zimbabwe and South Africa, which have gone some distance in setting up river basin organisations, it is important that integration becomes clearer at the conceptual level and is achieved in practice. In these countries river basin organisations which have been formed are tasked with, among other things, making river basin plans. A challenge which such organisations face is, what should be 'integrated' within river basin plans? This is particularly important in the context of managing water resources for livelihoods improvement. Furthermore, since IWRM is supposed to operate at multiple-scales it can also be asked, how can local water management practices and IWRM-based practices be integrated for improved livelihoods? These issues need to be addressed and clarified if IWRM is to contribute towards improving livelihoods.

3.4 Efforts at improving IWRM

Although opinion over the utility and conceptual soundness of IWRM is divided, it cannot be denied that the rise of the framework was in response to real problems in the water sector that needed solutions. IWRM emerged out of the realisation that there was need for change in approaches to water resources management, specifically in governance processes (Van der Zaag, 2005). One of the challenges which the water sector faced was that of fragmented policies in different government departments, and development initiatives which did not complement each other (Molle, 2008). Furthermore, there was little or no coordination between water management activities in the upstream and downstream parts of the same catchment, water quality issues were divorced from water quantity issues, and socio-economic issues, such as those concerning livelihoods, were not catered for in water resources management (*ibid*). Other challenges which also needed attention included those relating to how to deal with climate change and the challenge of attaining the Millennium Development Goals (MDGs) through water resources management (Van der Zaag, 2005). IWRM emerged out of this context as a normative framework, as a way of departing from the 'business-as-usual' approach to water resources management (*ibid*). Importantly, IWRM emerged to be a perspective through which water resources management problems could be analysed and solutions for which be sought (*ibid*). It can therefore be argued that although IWRM indeed does have weaknesses at the conceptual and operational level, it has at least served to expose the challenges of water resources management and offered possible solutions. Instead of doing away with IWRM altogether because of its weaknesses, real and perceived, it can be argued that what is needed is to improve on the framework.

Among the suggested ways of improving the framework is that instead of trying to implement 'full' IWRM developing countries should implement "Light IWRM', which is a scaled down version of IWRM. It has also been suggested that IWRM should incorporate elements of Integrated Natural Resources Management (INRM) so that it becomes more responsive to the needs of the poor.

Light IWRM: a practical IWRM?
In developing countries there have been attempts to implement a comprehensive list of integrated solutions to water resources management (Lankford *et al.*, 2007). Such attempts to set up the whole IWRM structure at once have more often than not failed. This failure is not surprising when one considers that even in rich western countries the perceived success of IWRM has been based on gradual adaptation of existing management structures and has been made possible by infrastructure investment made by private companies and financial support running into hundreds of millions of dollars (Lankford *et al.*, 2007). Such investments and financial resources are limited in developing countries. It has also been observed that countries which have adopted IWRM have abandoned projects, such as on water infrastructure development, which can deliver tangible benefits to water users. Such projects have become overshadowed by the need to reform the water sector and to put in place stakeholder participation platforms so as to comply with the tenets of IWRM (Shah *et al.*, 2005). The pitfall of continuing implementing IWRM in the manner in which it is currently being implemented in developing countries is that the needs of water users will not be addressed.

Awareness of the shortcomings identified above has led to calls for the development and implementation of 'Light IWRM' which is defined as IWRM which is opportunistic, adaptive and incremental in nature (Moriarty *et al.*, 2010). This type of IWRM is focused on sustainable service delivery, stops short of full integration, and respects political and institutional realities (Butterworth *et al.*, 2010; Warner *et al.*, 2010). 'Light IWRM' by virtue of focusing on service delivery also prioritises access to water for domestic and productive uses (Butterworth *et al.*, 2010). The strength of 'Light IWRM' is given as that it applies the core principles of IWRM at different water resources management levels, from the water user level to the river basin level, and also within the different sub-sectors (Butterworth *et al.*, 2010; Moriarty *et al.*, 2010). The approach therefore enables principles of IWRM to be built into different projects which would be the first step towards implementing full IWRM. The combined effects of such an approach is that it meets the needs of the water user (through service delivery) but does so without abandoning the tenets of IWRM such as stakeholder participation. 'Light IWRM' is different from IWRM 'proper' in that it pays more attention to service delivery rather than institutional and legislative interventions. Light IWRM is therefore more pragmatic and context-adapted in terms of its approaches and strategies. Light IWRM is service oriented and therefore within the framework priority is given to access to water and sanitation challenges and the development of irrigation services which address the livelihoods of water users at the local level.

'Light IWRM' appears to have the potential to solve the challenge of implementation capacity which most countries face, especially given the limited financial resources available to them. However, it is important to note that although proponents of 'Light IWRM' state that the approach retains the basic tenets of IWRM, such as integration and participation, the fact that within the approach it can be possible to pick and choose issues to attend to makes it potentially feasible to merge 'Light IWRM' with INRM and strategic water resources management.

IWRM and Integrated Natural Resources Management (INRM)

The low levels of socio-economic development and the high rates of environmental degradation in the developing world in general and in southern Africa in particular have caught the interest of both researchers and practitioners keen to find solutions to these challenges. This, to a large extent, accounts for the proliferation of research and development paradigms which try to offer a way out of under-development. IWRM and INRM are but two of developmental paradigms to emerge in the recent past offering solutions to challenges which underdeveloped southern Africa faces. Although IWRM emerged from the field of engineering and INRM from the agricultural sector, the two have a similar objective, which is that of improving land and water productivity (Twomlow *et al*., 2008). Improving land and water productivity can potentially improve food security which in turn improves livelihoods (*ibid*). This therefore calls for an understanding of how the two can be made complementary.

INRM is defined as, '*an approach to research that aims at improving livelihoods, system resilience, productivity and environmental services. In other words it aims to augment social, physical, human, natural and financial capital*' (CGIAR, 2001). INRM has also been defined as a process of incorporating the multiple aspects of natural resources use (biophysical, socio-political, or economic) into a system of sustainable management to meet multiple goals such as those of food security, profitability, risk aversion as well as goals of the wider community (like poverty alleviation, welfare of future generations, environmental conservation) (Campbell *et al*., 2002). It aims to help solve complex problems affecting natural resources management in agro-ecosystems and to safeguard natural resources and improve agricultural productivity. These goals are highly inter-dependent (Campbell *et al*., 2001). The use of INRM is perceived to make it possible to deal with issues such as the ones identified above because the approach empowers relevant stakeholders, makes it easy to resolve conflicting interests of stakeholders, fosters adaptive management capacity, and focuses on key causal elements and integrate levels of analysis (CGIAR, 2001). It can be added that some of these factors are also perceived to be the strength of IWRM. Another commonality with the IWRM framework is that INRM aims to integrate participatory research with community based natural resources management (CBNRM) (Twomlow *et al*., 2008). Furthermore, INRM goes beyond the traditional 'what' and 'where' questions to address the 'who' and 'how' aspects of social actors and processes (Gottre and White, 2001).

A major point to note, which distinguishes INRM from other approaches to resource management, is that INRM is being promoted with community groups and also with individual landowners (Lovell *et al.*, 2002). This means that it can be used in the management of communally owned, open access and privately owned natural resources. INRM makes use of multi-level analysis, livelihood, gender and community level analysis, policy, institutional and market analysis, and also analysis of natural resource status and dynamics (Thomas *et al.*, 2002). Hagmann *et al.*, (2002) state that the complexity of integration at different scales poses serious conceptual and organisational challenges, especially in relation to roles and mandates. It is important to note that INRM is considered as a problem solving approach which delivers tangible benefits.

From the above it is possible that both IWRM and INRM can operate along the same physical resource base and take a systems approach at multiple scales (Twomlow *et al.*, 2008). Both try to promote resource utilisation by taking into account how actions at one level can cascade to affect another level. However, a fundamental difference between the two is that INRM places the household at its centre while IWRM places the hydrological catchment as the basic unit of operation (*ibid*). In INRM therefore, the social boundary is the unit for intervention while in IWRM the physical boundary is the unit for intervention. An important lesson which IWRM could borrow from INRM include the recognition of the impact of the choice of scale of operation on livelihoods. Since INRM uses socially constructed scales as units for intervention, it is perceived as being more responsive to the livelihood needs of smallholder farmers. Importantly, if this is applied to the IWRM framework it can potentially result in emphasis being placed on the capital that households have being used to improve their livelihoods (Merrey *et al.*, 2005). However, it is also important to note that within the INRM framework the use of technologies to promote land productivity is encouraged, and some of these technologies include irrigation. Since IWRM is strong on hydrological assessments, a combination of INRM and IWRM can potentially benefit households since expertise from one field can contribute towards improving the type of interventions promoted by the other. Such synergies are not only important for the designing of irrigation-related technologies, but more importantly for the understanding of factors that influence smallholder farmer productivity and therefore contributing towards more holistic and practical solutions to the livelihoods of smallholder farmers.

Strategic and adaptive water resources management

Among the changes which the IWRM paradigm has brought to water resources management, and which has become the tenet of the management approach, is that of managing water resources along hydrological boundaries (Wegerich, 2009). This appears to have been strongly influenced by the perception that water is an integral part of the ecosystem, and therefore it is logical that the resource be managed along its natural boundaries than the human-created administrative ones (Warner *et al.*, 2008). While this approach has been widely promoted as part of the IWRM framework, it is not entirely new as managing water along hydrological boundaries dates as far back as the 18th century (Molle, 2009). In practice the management of water along hydrological boundaries has resulted in an institutional re-arrangement, for example, organisations managing water and stakeholder participation are now delineated along river basins rather than the traditional administrative boundaries. However, reality is that operating along hydrological boundaries ignores the administrative boundaries within which water users live, and within which socio-economic development is ordinarily planned for and implemented (Warner *et al.*, 2008). These and other criticisms have led to calls for a strategic approach to water resources management (Mollinga *et al.*, 2007). Strategic water resources management takes an 'issue' perspective instead of a watershed perspective (*ibid*). An issue-perspective, which is also known as a 'problemshed' approach, is different from a watershed perspective in that when identifying challenges that need to be dealt with, hydrological boundaries within which water resources are bound are ignored in favour of socio-economic or other boundaries within which water users live. Thus, for example, instead of mapping livelihood challenges following hydrological boundaries, which is normal practice within the IWRM framework, the problemshed does not confine such challenges to pre-defined boundaries. The perceived strength of this strategy is that it enables a water management issue to be approached as an open empirical question. This is opposed to the watershed approach in which issues boundaries are 'pre-defined spatially, sectorally and analytically through the primacy of water' (Mollinga *et al.*, 2007:707). Importantly, the problemshed approach enables an inter-disciplinary approach to solving issues within a geographical area, for example, by facilitating actors from outside the pre-defined stakeholder groups to be involved in contributing to addressing issues. It also makes it possible for problems to be tackled without hydrological boundaries defining which actors should be involved. An effect of using hydrological boundaries in water resources management has been that of dividing communities which are otherwise similar and share more or less the same challenges. Some challenges which rural communities face, for example, do not follow hydrological boundaries but socio-economically constructed boundaries, hence are difficult to solve within a watershed framework as advocated for within IWRM.

Among the criticisms levelled against the IWRM framework is that within the framework it has not been possible to respond to the needs of water users (Lankford *et al.*, 2007). Such needs, as mentioned earlier, include improved access to water, which can potentially improve livelihoods. The inability of water managers to respond to the needs of water users in river basins has been attributed to that the IWRM framework presents water managers with a long list of activities to execute, many of them simultaneously. Lankford *et al.*, (2007) suggest that instead of strictly adhering to the IWRM Principles, there is need to implement a water resources management model which focuses on problems on the ground. They refer to their suggested model as 'expedient water resources management.' They define expedient water resources management as 'advisable on practical rather than principled grounds-thus emphasising a shift towards problem identification and solution, and away from the adoption of accepted norms, including the Dublin Principles. An advantage of the adaptive-expedient approach is that it is a more practical framework which relies on designing activities against stated and relatively short-term goals of between five and ten years. They suggest that in implementing adaptive water resources management the following steps can be followed:
   1. Understanding and characterising the land, water, people and institutional behaviours in the basin

2. Establishing goals
3. Developing a management response to the goals
4. Generating activities that lead to, and drive steps 1-3

A fact to note is that an adaptive approach makes it possible for livelihood needs in specific communities to be identified, and therefore for practical action to be taken.

The foregoing sections have analysed some of the theoretical and operational controversies surrounding IWRM, and suggestions as to how the framework can be improved. While such controversies appear to be slowing down the implementation of IWRM, practical examples of how IWRM can be implemented in practice are beginning to emerge. What is particularly notable about some of the emerging practices is that they pay particular attention to the improvement of livelihoods. One example of such an approach is that of multiple use systems (MUS), which is presented in the next section.

3.5 An example of practical IWRM in southern Africa

Multiple use systems

As mentioned earlier, one of the most important challenges which most of southern Africa's rural households face is that of limited access to water as a physical resource. This compromises the health of households as they do not have enough water to meet their basic needs. Limited access to water also constrains the possibilities of households to improve their socio-economic status as their capacity to engage in productive activities is reduced. To address such needs, in the past projects characteristically tackled single aspects, for example, providing water supply systems for domestic needs only, or providing infrastructure for irrigation only. Interestingly, the MDGs and the World Summit on Sustainable Development (WSSD) make explicit reference to water for drinking and sanitation but little or no reference at all to water for productive uses (Merrey *et al.*, 2005). While 'single objective' projects have their merits, it has increasingly become apparent that livelihood challenges need to be tackled from a more holistic approach, that is one which addresses multiple needs of water users. The multiple water needs of households, which basically fall into water for domestic uses and water for productive uses, can be met through multiple use systems (MUS) (Van Koppen, *et al.*, 2006).

Multiple use systems are designed to provide households with water (in large enough quantities) to meet their domestic and productive uses. Thus instead of designing a water supply system which only provides enough water for domestic uses, enough water to irrigate small gardens or to engage in micro-industries should also be provided by the same system. Productive uses of water can potentially enable households to increase household income and therefore to improve their livelihoods (Van Koppen *et al.*, 2006).

In the context of southern Africa the attraction of MUS is that the approach has the promise of not only solving the challenge of access to water for domestic needs, which is critical for the well-being of households, but also for enabling the same households to increase their income and therefore their livelihoods. This in a practical way integrates water resources management with practical livelihood issues. Proponents of MUS also argue that in some cases the poor have limited access to land, which limits their access to irrigation schemes (Moriarty *et al.*, 2006). Thus in areas where water infrastructure development targets irrigation development alone, poor households are likely not to benefit from such schemes. As a result, schemes which target the provision of water above domestic needs are more likely to reach the poor and potentially improve their livelihoods. The approach potentially can complement IWRM, which also seeks to integrate different issues for socio-economic development.

Evidence from Zimbabwe's Runde catchment, for example, suggests that multiple use systems are viable and do improve livelihoods. Promoting productive water use through community gardens set up at high yielding water points enabled households to increase their incomes significantly (Robinson *et al.*, 2004). High yielding water points are considered to be those which provide more water than is needed for domestic uses of the community it serves. Furthermore, such approaches have also been found to be more effective at reaching out to poorer members of the community than traditional approaches of setting up irrigation schemes (*ibid*). Such evidence has been used to promote the upgrading of family wells by deepening them so that they yield enough water for both domestic and productive uses.

3.6 Discussion and conclusions

It is more than a decade since some countries in southern Africa embarked on the implementation of IWRM. The time which has passed allows for reflection on experiences of implementing IWRM in the region. This discussion analyses the IWRM in relation to livelihoods, with a major focus on the concept and practice of IWRM and the context within which the framework is being implemented. However, from the onset it needs to be emphasised that through water resources management, in one way or another, some form of socio-economic development has been delivered. In the hydraulic mission era (Allan, 2003) the main objective of water resources management was the development of water infrastructure. Such developments in turn increased, for example, power generation capacities which in turn benefitted industrial production. The downstream impact was employment creation and in that way livelihoods benefitted. However, it has become apparent that poverty and other livelihood challenges have to be tackled in a more direct manner. This is exemplified, for example, by the formulation of the Millennium Development Goals (MDGs) which target reducing the proportion of population living below a certain standard. In line with such thinking, it is therefore necessary that resource management addresses livelihood challenges in a more direct way rather than making an indirect contribution which might have less impact or take longer to achieve. It is in this context that the analysis of IWRM focuses on the possibilities of the framework to contribute towards livelihoods.

There is little doubt that in the recent past IWRM has attracted enormous attention from policy-makers, development practitioners and academics. Examples and cases from around the world have been used to analyse IWRM as a concept and to evaluate the challenges of implementing the framework. This has resulted, for example, in a proliferation of definitions of IWRM, explanations on how IWRM can be implemented and the development of theories surrounding water resources management. Interest in IWRM at the global level is evident in that annually a number of conferences and symposia are held to discuss IWRM. Among them are the World Water Week which is held in Stockholm, Sweden, and the WaterNet/GWP/Warfsa Symposium which is held in southern Africa each year. In addition to these, the Global Water Partnership and the World Water Council represent organisations dedicated to the promotion of IWRM across the globe. The IWRM agenda has also been adopted by powerful global actors, such as the United Nations and its key bodies such the World Bank, the Food and Agricultural Organisation and the United Nations Educational, Scientific and Cultural Organisation (UNESCO). However, the sad reality is that while these organisations have devoted immense resources and time to promoting and debating IWRM, at the local level little has changed as far as improved access to water is concerned. The majority of the poor continue to suffer from limited access to water for both domestic and productive uses. Livelihood needs which can be solved through improved access to water, such as improved household food security which can be achieved through improved crop productivity, remain challenged. This perhaps is the clearest indictment against IWRM and continuing with its implementation in its current format.

It has become clear that some of the challenges of implementing IWRM stem from the conceptually unclear principles of the framework which do not adequately guide the practitioner (Biswas, 2004). Sceptics have argued that at a conceptual level IWRM can be improved by, for example, explicitly stating that the framework is about the improvement of livelihoods (Merrey et al., 2005). They argue that such emphasis can focus the priorities of water resources management. Another issue which transcends both the conceptual and practical level is that of what to integrate within the framework. There is no clarity on what needs to be integrated within the socio-economic system. This lack of clarity has livelihoods implications. Furthermore, although within the IWRM framework sectoral integration is supposed to be implemented, it is not clear which sectors are supposed to be roped into water resources management and how that can be achieved. Sceptics also argue that it is futile to try to integrate because there is no evidence that integration does work (Dent, 2011). Calls against integration on these grounds are rather puzzling because the fact that integration does or does not work can only be obtained after it has been attempted (ibid). Thus one can argue that lack of evidence that integration does work should not be used as evidence that integration does not work. However, it can be argued that clarifying what IWRM is about, or ought to be, might not necessarily translate to clear operational guidelines as far as its implementation is concerned. The MDGs, for example, clearly articulate developmental goals to be achieved but the sad reality is that most developing countries are in danger of missing the set targets. Furthermore, there is evidence that even among sceptics there is no agreement on what should be integrated within IWRM. Others are of the opinion that IWRM tries to integrate too many things which ends up affecting implementation. Yet others argue that critical resources such as forest resources and biodiversity are not integrated into IWRM although they are critical for livelihoods. It can therefore be concluded that there is little possibility that consensus can be reached over what exactly IWRM should integrate. Energy should therefore be directed towards taking practical steps to provide the poor with water. That can potentially trigger processes which can lead to the improvement of livelihoods.

What is encouraging, from a livelihoods perspective, is that some have begun to advocate for practical guidelines that can be implemented and potentially improve livelihoods. The integration of domestic and productive water supply systems, which is referred to multiple use systems, is one such example (Moriarty et al., 2004). Proponents of MUS suggest that when water infrastructure is being provided, the target should be to provide more water than is necessary for domestic consumption alone. This guides practical action because for a given community it is possible, using population and per capita consumption estimates, to calculate domestic water needs. To this estimate provision can then be made for water to meet productive uses. This can potentially lead to increased food availability and income at the household level through gardening or other income generating activities. MUS as a practical suggestion stands out in that it integrates perspectives from other fields, such as rural sociology and development studies which have proven that if households have improved access to water it is possible for them to embark on micro-enterprises and improve household income and hence improve their livelihoods. Interestingly, calls for IWRM to be informed by INRM recognise the need for integrated approaches to improving rural livelihoods. Such an approach can potentially merge the social system in which smallholder farmers, who are the water users at the local level, with the natural system, in which water and other physical resources exists.

Although the challenges of implementing IWRM in developing countries are real and cannot be ignored, it is also fair to suggest that some of the identified challenges are not internal to the framework but are related to the wider socio-political environment, that is the context within which IWRM is implemented. Some of the evidence which has been used to bolster the argument that IWRM is difficult to implement has come from the experiences of developing countries, such as Zimbabwe and South Africa. These countries, which were among the first to embrace IWRM in southern Africa, appear to be struggling to implement IWRM. In South Africa, for example, more than a decade after the water reforms were passed, there are less than a handful of river basin organisations in place. While this has been taken to suggest that it is difficult to implement IWRM, it is important to separate implementation challenges internal to IWRM from those emanating from outside the framework. In South Africa for example, the formation of river basin organisation has been slowed down by, among other things, resistance to the change in status quo by commercial white farmers. This has been witnessed in cases where there have been attempts to transform irrigation boards to multi-stakeholder platforms. Such transformation can potentially improve livelihoods as the interests of blacks will be considered in water resources management. However, the transformation by implication would result in loss of privileges in access to water by the commercial farmers. It can therefore be argued that in such cases the implementation of IWRM is being made difficult by ingrained social attitudes and does not reflect the weaknesses of IWRM. Furthermore, in most developing countries motives for implementing IWRM have not been wholly local, but the reforms have been driven by interests of western countries (Swatuk, 2005). Thus in some countries there have been half-hearted attempts at implementing IWRM for purposes of accessing donor funds or conforming with global trends (Petit and Baron, 2009). This can explain the reluctance to implement the water sector reforms.

Addressing the needs of the poor is urgent. In southern Africa the majority of the poor reside in rural areas and are dependent on rainfed agriculture. In river basins such as the Limpopo, poverty is a result of multiple causes, some which can be attributed to natural factors and others to human factors. IWRM, which has become the dominant water resources management framework in the region (Swatuk, 2005) has been portrayed as being capable of maximising socio-economic development in a sustainable manner (GWP, 2000). This has raised hopes that livelihoods can be improved by managing water within the framework. However, if it is acknowledged that under-development has multiple causes, of which limited access to water is one, then it should logically follow that no single sector can solve livelihood challenges in the region, no matter how enticing the promises articulated in favour of a single solution are. Since the causes of poverty vary across space and time, it is therefore true that potential solutions to the problem of poverty cannot be uniform across the globe, and across different eras. On this understanding it can therefore be argued that part of the problem with IWRM is that within the framework generic solutions to water resources management problems are offered. This explains why, for example, some of the Dublin Principles, such as that water should be treated as an economic good, appear to be misplaced in the context of rural southern Africa. A generic approach is therefore the weakest challenge of IWRM.

It should be appreciated that at least the water sector has openly campaigned for integrated socio-economic development. An integrated approach offers the best hope for tackling livelihood challenges, but the question remains, what needs to be integrated and how can integration be achieved? However, one can argue that for water resources management to make meaningful contribution towards improved livelihoods there is need to better understand how water is managed at the local level.

Chapter 4

Dynamics and complexities of practices in local water management in Zimbabwe in the IWRM era

4. 1 Introduction

In southern Africa IWRM's ascendance was demonstrated by water sector reforms which were largely based on IWRM principles (Manzungu, 2004; Swatuk, 2005). Zimbabwe was among the first countries in southern Africa to embrace IWRM-led water reforms. In 1998 the country promulgated pro-IWRM water legislation, namely the Water Act (Zimbabwe, 1998a) and the Zimbabwe National Water Authority Act (Zimbabwe, 1998b) which were supported by an IWRM policy (MRRWD, n.d.).[5] Among the changes brought by the reforms was the introduction of hydrological-based water management institutions in the form of catchment and sub-catchment councils. These replaced institutions that were based on administrative boundaries and were meant to bring organisations managing water resources closer to the water users. Catchment and subcatchment councils were also to act as stakeholder platforms. Given that these IWRM-based institutions were created to improve water resources management in the country, it can be argued that they represent the latest intervention in water resources management in Zimbabwe. However, as Long (2001) observes, although interventions seek to change how things are done, the same interventions are affected by the context in which they are implemented. This makes it imperative to understand the context in which the intervention is being implemented. The main objective of this chapter is to analyse practices in water resources management at the local level. The following research questions are asked to guide analysis: *what drives practices in water resources managed at the local level?* The sub-questions which the chapter further asks are: *how is water accessed and water infrastructure maintained at the local level? What characterises water user dynamics at the local level, and what influences these dynamics?* Answering these questions will shed light on how water resources are being managed at the local level in reality. That in turn will contribute towards understanding whether, or not, IWRM as an intervention in Zimbabwe's water sector has affected local water management at the local level.

This chapter is based on data collected in Ward 1 of the Insiza district, which is located in southwest Zimbabwe in the Matabeleland South province. The ward falls under the Upper Mzingwane subcatchment, which is a part of the Mzingwane Catchment Council that has water resource management jurisdiction of the entire Limpopo river basin in Zimbabwe (see Fig. 2.1 in Chapter 2). It also falls under Chief Ndube in the Glass Block

---

[5] However, in 2006 a major policy shift in the water sector occurred when the government ordered the national water authority to take over urban water supply (see chapter 7).

communal area. Glass Block is one of the three communal areas in the district. Data collection, carried out between March 2006 and February 2009, included interviews with key informants, focus group discussions and observations of water-related activities and events. Key informants included community leaders such as village heads, the ward councillor and members of water point committees and committees managing small gardens. Other key informants included village pump minders and civil servants such as those in the Ministry of Health and the District Development Fund (DDF)[6]. Focus group discussions were held with water users who fell in such categories as domestic water users, livestock owners, irrigators and small gardeners. It is important to underline that in practice these users are not distinct and separate groups.

The chapter is structured as outlined below. The empirical material is preceded by the analytical focus of the study, which is presented in section 4.2. Thereafter is a literature review of local water management in Zimbabwe. This is followed by a description and analysis of local water management practices in the study area. Three cases are used to illustrate the range of issues that are involved in local water management. The first two cases involve a borehole, which is a common type of water infrastructure found in many rural areas of Zimbabwe. Boreholes enable rural people to exploit groundwater. Most rural communities depend on boreholes for their domestic water needs. The first case involves a borehole[7] fitted with a hand pump, which is common in many communal areas of Zimbabwe, while the second case involves a borehole fitted with a windmill. The third case involves a wetland, which is another common source of water for rural people in Zimbabwe. Wetlands are a source of both domestic and productive water, and are critical to the livelihoods of many rural households in Zimbabwe (Owen *et al.*, 1995). The chapter ends with a synopsis of local water management in Zimbabwe in the light of the empirical evidence.

4.2. Analytical focus of the study

When discussing local water management it is important to start by looking at how it is conceptualised (Fatch *et al.*, 2010). In this regard the subsidiarity principle, which is one of the principles in the United Nation's Agenda 21, is relevant. The principle states that water should be managed at the lowest appropriate level. Although the principle has been widely welcomed in the water resource management discourse, it has not clarified what 'local' refers to. Giddens (1983) opts for the 'locale' instead of 'place' on the basis that place connotes a point in space. 'Locale' encompasses the social systems within the geographic location. This chapter leans more towards the ideas of Giddens. The local is pertinent given that water management, within IWRM, is supposed to be based on hydrological scales, which has seen the creation of new institutions such as the

---

[6] The DDF is a government department which was set up mainly to help Rural District Councils to develop infrastructure.

[7] A borehole is a machine-drilled hole into the ground from which underground water is tapped. Boreholes can be as deep as 60m. Water can be pumped from boreholes using different types of pumps, such as electric pumps, hand pumps, wind-driven pumps (windmills) or other types of pumps.

subcatchment and catchment councils in Zimbabwe, and Catchment Management Agencies in South Africa. Given that hydrologically-defined areas can have a very large range, from the plot level and the micro-watershed to entire river basins, the question then becomes how to actualise this hydrologically-defined local. In Zimbabwe this has been most apparent when it comes to organising stakeholder representation and participation in water management (Latham, 2002; Kujinga and Jonker, 2006; Mabiza *et al.*, 2006).

Sneddon *et al.* (2002) argue that there is no point in using hydrologically-based boundaries to define local water management because of the multiple sub-systems that are nested within other hydrologically-based systems. Thus each hydrological level can be sub-divided and constitute a 'local' level in its own right. Besides, due to human activities such as inter-basin transfers, which incorporate areas outside those marked by the natural watershed boundaries, hydrological boundaries are no longer that clear-cut (*ibid*). The hydrologically-defined 'local' is also problematic because it does not exist in the lived-in social, economic and political reality in which the water users are enmeshed (Merrey, 2009). Thus to neglect socio-politico-administrative units on the basis that water does not respect administrative boundaries may be counterproductive because these are ubiquitous in the rural landscape and are the locus of many development projects and programmes. It is also notable that typically most planning processes occur or consider these socio-political administrative units. It is however, important to note that deficiencies of the hydrologically constructed 'local' must not be misconstrued to mean that the lived-in socio-politically (administrative) demarcated boundaries clearly define 'local' in terms of what it is, its size, and how and where it best fits with the water management hierarchy. This is because the socio-politico-administrative 'local' is, in many cases, laden with conceptual and practical problems.[8]

The interest in this chapter is to understand local water management as it happens. Consequently I apply the concept of social practice. Reckwitz (2002:249) defines practice(s) as *"...routinized type of behaviour which consists of several elements, interconnected to one other: forms of bodily activities, forms of mental activities, 'things' and their use, a background knowledge in the form of understanding, know-how, states of emotion and motivational knowledge."* In this chapter attention is paid to the relations between agents (local people) and objects (water, technologies) as local water management is a function of both social and the material dimensions (Van der Zaag, 1992). Although Reckwitz argues that practices do not imply social interaction, I consider them to be closely related. This is because in a situation where sources of water are not privately owned but are shared, social interaction is inevitable. Interactions refer to the processes which ensue when people come together and exchange goods, words or share experiences (Van der Zaag, 1992: 5). Their documentation and analysis may help explain how and why water is managed in a particular manner. Of interest here is how water

---

[8] For example in the Zimbabwean rural areas there are 'traditional villages' which can be loosely defined as a number of households under a village head, state -defined 'villages' comprised of several traditional villages, whose leaders constitute the Village Development Committee (VIDCO). For planning purposes a ward, which is made up of about six VIDCOs, can be said to be the local unit as it is the smallest planning unit in the country.

users interact over a range of issues, such as access to water, as well as repair and maintenance of infrastructure.

Related to the above is that water management, whether it is at the local or a higher level, involves some degree of cooperation and conflict simply because water user interactions are rarely neutral. Social behaviour between two or more individuals is cooperative if the behaviour of each is valuable to the other, or to a third person (Ramirez-Sanchez (2006) while conflict is a relationship between two or more parties who perceive that they have incompatible or competing goals or means of achieving those goals (Schmid, 1998). Zeitoun and Mirumachi (2008), however, reject the binary notion of conflict or cooperation on the basis that these represent different gradations of what essentially is the same relationship, namely social-interaction. They argue that conflict is often wrongly perceived as a negative relationship or outcome, although it can have positive outcomes. For example, parties which are in conflict may realise that they need to co-operate in order to secure their own interests. In this regard long held beliefs that water scarcity is the principal cause of conflicts may not hold.

Issues that relate to social practices, interactions, conflict and co-operation in water management all happen within one institutional setting or another. Institutions have been described as composed of rules of the game (North, 1990). Cleaver (2000, 2002) rejects the rational choice perspective to institutions, opting instead for an approach which embraces social and historical factors to explain the nature of institutions. She uses the concept of institutional bricolage to show that institutions in natural resources management are partial, changeable and evolving and borrow from existing styles social relations. Unlike formal institutions which tend to be set up for a specific mandate, institutions operating at the local level tend to be more holistic in terms of the issues they deal with. Often they involve the same social actors. This is at variance with normative models which suggest that institutions with 'clear roles and responsibilities' can be designed (Ostrom, 1990).

The following section presents an overview of sources of water and the uses of water in the study area. This is followed by a detailed analysis of water user relations at three sources of water in the ward as described above. The discussion centres on how water users gain access to sources of water and the nature of the interactions among the different water user groups.

4.3 Local water management in Zimbabwe: an overview

Literature on local water management in Zimbabwe can broadly be placed into two categories. The first category deals with how water that is used for productive purposes (agriculture) is managed while the second category focuses on domestic water supply. In a way this reflects the disciplinary distinctions that informed water resource management before IWRM. To what degree IWRM legislation and policy has changed the disciplinary separation remains a point of discussion.

As far as the first category is concerned one can observe that there have been numerous studies dealing directly and indirectly with the topic of local water management in the last two decades[9]. Over time the focus shifted from irrigation to water resources management. As far as irrigation is concerned the discourse on local water management revolved around smallholder irrigation schemes with a bias towards group irrigation schemes where collective action issues apply. Evidence is presented to show that irrigation was practiced by indigenous people prior to colonisation in 1890 and has continued in one form or another (Bolding et al., 1999). With time the colonial state got interested in the smallholder irrigation schemes and in 1913 the government launched the development of smallholder schemes (Rukuni, 1988). These schemes were conceived as a means of famine relief. Later they were seen as a device to settle black farmers dispossessed of their land as part and parcel of the colonisation process. By 1928 the colonial government was actively involved in the provision of services to the smallholder farmers. In these schemes the role of smallholder farmers was significantly reduced (ibid). With time national level politics resulted in government's benign benevolence changing into open dominance of farmers as the government wrestled with how to define a role for smallholder irrigation schemes. As farmers were turned into mere tenants local water management suffered considerable harm (Manzungu, 1999; Magadlela, 2000). Manzungu and Machiridza (2005) show that after independence the post-colonial state more or less continued with colonial policy despite protestations of local people.

Technical studies tended to dominate the debate on smallholder irrigation in the first two decades after independence (Donkor, 1991; Makadho, 1996). Other studies took a broader perspective and investigated the role of technical and social factors in local irrigation management, an example of which is the collection of papers in Manzungu and van der Zaag (1996). Both technical and social factors were shown to play an important role in enhancing or constraining the performance of smallholder irrigation schemes (Chidenga, 2003). Just as was the case during the colonial era national politics were demonstrated to affect local water management. Zawe's (2006) study shows how (fast track) land reform impacted on irrigation development, and illustrated the need for balance to be struck between political and socio-economic realities. Although these various studies differed in terms of focus and hailed from different disciplines they nevertheless reach the general conclusion that farmer management was better than state-led management.

As IWRM became widely accepted in water management the focus of debates dealing with local water management also changed. In the aftermath of the 1998 legislation there was interest in how local people participated in the new water management institutions (Latham, 2003; Mtisi, 2008). The question of representation vis-à-vis how 'legitimate' were representatives of local level interests and whether local people could effectively assert their interests received critical discussion against a backdrop of many institutional layers (Sithole, 2001; Kujinga and Manzungu; 2004; Chereni, 2007). Evidence

---

[9] Since the 1990s some 10 PhD and numerous MSc studies have studied the subject of water resource management at the local level in Zimbabwe. PhD studies include those by Donkor (1991), Makadho (1994), Vijfhuizen (1998), Manzungu (1999), Magadlela (2000), Chidenga (2003), Bolding, (2004), Zawe (2006) and Mtisi (2008).

challenged the assumption that devolution guarantees participation and showed that the interactions between and among different stakeholders were defined by power.

As far as domestic water supply was concerned there was a focus on the ability and capacity of local people to manage water sources and related infrastructure (Machingambi and Manzungu, 2003). Makoni *et al.*, (2004) found that in rural areas women were more active in the use and management of water than men. In the same rural areas there was a tendency to segment water services, such as putting up facilities strictly for domestic water supply or crop production only (Katsi *et al.*, 2007). Robinson *et al* (2004) observe that in most cases shallow wells, deep wells fitted with bush pumps or rope and washer pumps were provided to meet rural water needs. Small dams provided water for crop irrigation. This was seen as disadvantaging rural water users who also need water for productive uses to meet their livelihood needs. Although the Zimbabwean IWRM policy reaffirms one of the IWRM principles of treating water as an economic good (MRRWD, n.d.), there is, however, no evidence to show that this policy was systematically implemented. What seems to have happened is that the government, in the aftermath of the water reforms, supposedly in the spirit of decentralising the management of water, pushed the financing of water sources to the local level. Against this background Derman *et al.*, (2005) raised the question of access to water and argued that a human right approach to water was more in keeping with the spirit of the water reforms, and pointed out that this was culturally accepted at the local level.

Having analysed some of the trends observed in local water management at various points in time, the following sections move on to analyse practices in local water management in the Upper Mzingwane subcatchment in the IWRM-era.

4.4 Water management in Ward 1

4.4.1 Overview of water sources and water uses

The individual cases that are described below are preceded by an overview of the sources of water in the ward, focusing on the types of water sources and infrastructure, their uses and the rules which apply to them. The overview sets the scene for detailed analysis of how water users engage each other over the management of water. Fig. 4.1 shows the location of the study area while Fig. 4. 2 shows the location of various water sources that are found in the ward, which include small dams, a wetland (*dambo*), shallow wells, hand pump-operated boreholes and a wind-powered borehole.

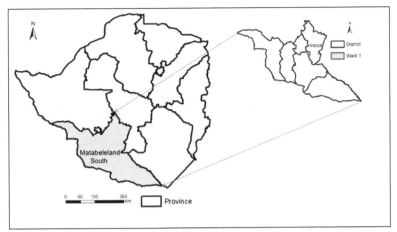

Figure 4.1: Location of the study area

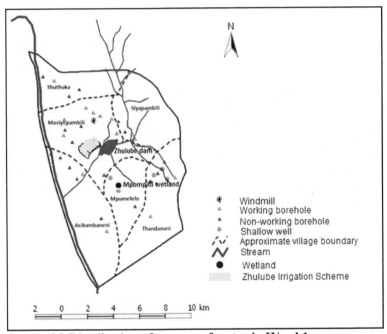

Figure 4.2 Distribution of sources of water in Ward 1

Many of the small dams in the ward were constructed by the District Development Fund (DDF). There are, however, others that were a result of initiatives by non-state actors such as faith-based organisations. For example World Vision, an international Non-Governmental Organisation (NGO), funded the construction of the 800 000m³ Zhulube Dam which supplies water to the Zhulube Irrigation Scheme. Most of the wells in the ward are shallow, being on average between 2-5m deep and about 1.5m in diameter[10]. Some of these wells have had hand pumps installed on them so as to reduce the burden of lifting water. Water users in the ward commonly refer to such 'improved/modified' wells as boreholes. The most common type of hand pump is the B-Type Bush pump[11]. In this chapter such waterpoints are referred to as boreholes in order to avoid confusing them with shallow wells which have not been fitted with hand pumps, and to be consistent with the terminology used in the ward. The installation of a hand pump changes the dynamics of managing a waterpoint since the pump needs maintenance. Apart from these 'improved/modified wells' there are also a number of machine drilled boreholes. Generally the water sources in the ward are used communally.

Water is mainly used for domestic purposes, livestock watering, irrigating crops, brick making and for processing gold ore. Water for laundry and bathing purposes is obtained from sources which are considered to contain water which is not of potable quality, such as small dams. Water for cooking and drinking is fetched from sources which are considered to have water of potable quality. Such sources include boreholes, shallow wells and the wetland. In almost all cases each water source was used for more than one water use[12]. What tended to vary were the priorities of water use.

Water quantity and water quality determined use of a water source. As already highlighted above, sources which were considered to have water of a potable quality were generally reserved for domestic uses. In the dry season, especially between August and October, such sources were strictly reserved for domestic water uses ahead of such uses as livestock watering. Uses which did not necessarily require water of potable quality but needed large quantities of water, for example, brick making and gold panning, were catered for from small dams, streams and rivers. Table 4.1 gives a summary of the views of the water users regarding the reliability of water sources. In the case of boreholes with hand pumps and the windmill, reliability was determined by the frequency of breakdowns. Only the windmill and the wetland were considered to be very reliable because at these waterpoints water was available throughout the year.

---

[10]The diameter is a result of the fact that the wells are dug using hand-held tools which means there is need for enough room for the person digging to swing the tools around.

[11] This is a manually operated Zimbabwean made pump which was developed especially for use in the rural areas

[12]This is consistent with the findings of Van Koppen et al., (2006) who found that local people use water from one source for multiple purposes. There have been calls to consider this fact in the designing of rural water schemes. 'Multiple-use systems' (MUS) are seen as critical to improve access to safe water and poverty alleviation.

Table 4.1: Reliability of sources of water in Ward 1, Insiza District

Source	Reliability	Reason(s)
Boreholes	Medium	Some dry up in winter Break down often
Small dams	Medium	Dry up in winter Provide adequate water for livestock watering
Windmill	High	Rarely breaks down Provides adequate water for irrigation
Wetland	High	Provides adequate water all year round
Shallow wells	Low	Dry up in winter Yield low amounts of water all year round
Rivers and streams	Low	Dry up in winter

4.4.2 Examples of local water management

This section analyses three cases which represent different aspects of local water resources management in Ward 1. As mentioned earlier, the cases are based on a shallow well with a hand pump, a borehole fitted with a windmill to pump water, and a wetland. Fig. 4.3 shows the location of the waterpoints that will be analysed. An important thing to note is that the three cases are within a 10 km radius.

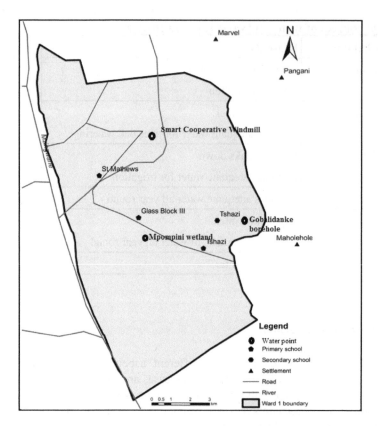

Figure 4.3. Location of the Smart Cooperative Garden windmill, the Gobalidanke borehole and the Mpompini wetland in Ward 1, Insiza District

Case 1: Common source, divergent behaviour: water use and users at the Gobalidanke borehole

The Gobalidanke borehole is actually a shallow well fitted with a hand pump (see earlier explanation on distinction between shallow wells and boreholes). From around September until late October (when summer temperatures peak and sources of water such as shallow wells dry up) water users converge at the borehole sometimes from as early as 4 a.m. until around 6 p.m. In 2007, at the height of a drought, I monitored water use at the borehole daily over a 7 day period. The daily average number of users who came to fetch water was 115. The highest number was 149 while the lowest was 95. In 2008 although the number of users significantly reduced the pattern for water use remained the same. This was probably because both years had a similar rainfall pattern. Fig. 4.4 shows water use pattern at the water point in October 2008 averaged over a period of 7 consecutive days. The highest number of water users, and the amount of water fetched from the borehole, was recorded between 7 and 8 am. This coincided with the time households fetched water before embarking on their daily household chores or work in the fields. The decline in numbers from then onwards coincides with the engagements in the fields.

From 3 p.m. there is a gradual rise in the number of water users coming to the borehole when people troop from their daily activities to the borehole. Figures 4.5 and 4.6 show activity at the borehole.

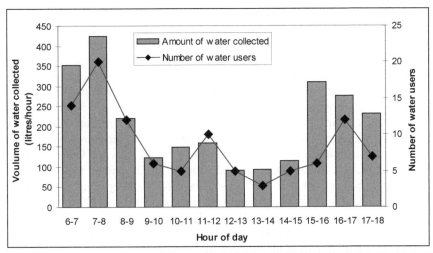

Figure 4.4. Averaged water use pattern at Gobalidanke borehole, October 2008

Figure 4.5: Water users fetch water from the broken borehole

Figure 4.6: Water users prepare to leave the Gobalidanke borehole after fetching water

The borehole is used by water users from Wards 1 and 2. It is used by about 39 households from Thandanani Village (Ward 1), and households from three villages in Ward 2. It is almost equidistant between Thandanani village and the villages under village heads Shwadi, Jabulani and Bachipa's who hail from Ward 2. Its location near a road which joins the Mahole Business Centre to the Zhulube Business Centre means that passers-by also access it. There is also a clinic located just 40 meters away. Although the clinic is connected to a piped water system, which represents its main source of water, in times of water shortage, the nurses from the clinic get water for their domestic needs from the borehole. The majority of the regular users, however, are villagers from Wards 1 and 2. Water users from Ward 2 also have access to a piped water supply system. The

55

piped water supply system was part of a government programme aimed at improving water supply to health centres in the country, and the villages benefitted from the same scheme by virtue of close proximity to the intended target of the scheme.

Access rights to the borehole, which are contested by users from the two wards, are influenced by beliefs, location of the borehole, and investment by users. Water users from Thandanani Village (Ward 1) feel that since the borehole is located in their village they have the right to the borehole. According to them, they are entitled to exclude water users from other wards. They also argue that water users from Ward 2 are privileged since they have access to a piped water supply system. However, water users from Ward 2 also claim access rights to the borehole. They claim that between 1976 and 1977, before the creation of the existing ward boundaries, it was they, working together with a land development officer, who took the initiative to deepen and protect the *umtombo* (well) at Gobalidanke. After independence it was converted into a borehole by the United Nations Children's Fund (UNICEF). Village head Shwadi from Ward 2 claimed that only three people from Ward 1 had joined men from his village in the deepening of the well. Because of this investment in hydraulic property (Diemer and Huibers, 1996), the village head was of the opinion that people from Ward 1 did not have the right to deny water users from Ward 2 access to the borehole. He also pointed out that the site where the borehole is plays an important part in the rain making ceremonies.

Events of the recent past have also served to heighten the contestations over access to the borehole. From around 2000 when the country experienced a series of droughts, NGOs intervened by distributing food. One such NGO, World Vision, started a 'food-for-assets' programme under which it gave food hand-outs to households who worked on community projects. One such project was the rehabilitation of the Gobalidanke gully. However, the project involved only households from Ward 1, which angered people from Ward 2. This was worsened by the fact that World Vision also funded an irrigation scheme in Ward 1 and only households from the ward became members in the scheme. People from Ward 2 argue that since they did not benefit from the food aid programme or from other developmental schemes as Ward 1 has, they have the right to use Gobalidanke in any manner they want to.

The following are some of the important rules that apply to the water source: i) users are not allowed to fetch water in 200 litre drums, ii) laundry should be done at designated points within the vicinity of borehole, iii) children are not allowed to fetch water without adult supervision or play near the borehole, and iv) the pump handle should not be banged on the ground.

However, the rules were not strictly observed. Rules concerning where laundry is done and the amount of water which users are allowed to fetch at any one time are commonly broken. It is common to see women from both wards doing laundry in places very close to the borehole. Dirty water from the washing of clothes formed filthy puddles around the borehole. Conveniently offenders claimed not to be aware of the rule prohibiting doing laundry within close proximity to the borehole.

During the time of the research the handpump at the borehole was broken down. As mentioned earlier, some of the boreholes in the ward are actually shallow wells which have been improved through deepening and the installation of a handpump. When such boreholes break down water users simply remove the top part of the handpump and the break the concrete covering the mouth of the well so that they can continue to fetch water. Such was the case at Gobalidanke. In the absence of a functional committee it has not been possible for the users to organise themselves to make financial contributions for its repair. Although challenges relating to collection of money for repair and maintenance are common throughout the ward, what makes the problems at Gobalidanke stand out is that there are no institutional structures through which water users can be organised. This 'institutional vacuum' became apparent when it was found that only one member of the waterpoint committee was still living in the area. However, even before the other committee members had left the area, institutional problems at the borehole were already showing. The underlying problem could have been the fact that waterpoint committee members came from water users from Thandanani Village in Ward 1. Water users from Ward 2 were not included in the committee as they were considered by those from Ward 1 not to be legitimate water users of the borehole. Water users from Ward 2 claimed ignorance about rules operating at the borehole. In some cases, however, they openly defied the waterpoint committee. This was because they knew that no reprisals would befall them.

The management problems at the borehole have not attracted the intervention of higher authorities. It appeared the problems were still considered 'local' and not serious enough to warrant the attention of the Chief, the Rural District Council (RDC) or councillors. Relevant village heads have not been able to come together to try and solve the issues. The stated reason was that the village heads hail from different wards. Also there is no one village head with the power to summon the other to a meeting (although such meetings can be held by mutual consent). This has also been influenced by that the village head who has tried to call for the meeting is a woman. Village heads from Ward 2 were said to have simply refused to come for meetings called for by village head Nonto from Thandanani Village (Ward 1). The situation is not helped by the fact that the NGO that is active in the area, World Vision, operates in Ward 1 and not in Ward 2 as already noted.

Case 2: Interconnectedness of institutions: case of the Smart Cooperative windmill

The Smart Cooperative Garden began in 1990 as an initiative of the Mother's Union of the Anglican Church in Ward 1. The cooperative has 23 members. Of these 8 are affiliated to the Anglican Church while the rest are members of other churches such as the Seventh Day Adventist, Zion and the Apostolic Churches. Most of the members of the garden (20) are from Mpumelelo Village. Only 3 are from a neighbouring village of Thuthuka. Three of the members are also village heads.

The cooperative uses a windmill to pump water from a borehole into a storage tank which has a capacity of about 54 000 litres. The garden is about 0.6 ha in size. Each member of the cooperative has 18 beds, each measuring 5 m long and 2 m wide. Vegetables such as cabbages are grown throughout the year. However, in winter, more space is devoted to growing vegetables, peas, carrots and some maize. Figure 4.7 shows the windmill at the Smart Cooperative Garden.

Figure 4.7 The windmill at the Smart Cooperative Garden

There is a management committee in place, which is made up of four members, namely the chairman, the treasurer, the secretary and an additional member. Kinship, religious and social ties bind the committee members. The chairman and the secretary are husband and wife. The chairman is also a sub-deacon in the Anglican Church. The secretary is the treasurer in the same church and also a member of the choir. This web of relationships was said to be the basis on which the nutritional garden was established. Interestingly the committee was said to have never been changed since it was formed. Members of the nutritional garden expressed satisfaction with the committee's composition and performance.

As with other water infrastructure, the windmill requires repair and maintenance measures to keep it working. It was claimed that in the more than 20 years in which the windmill has been running it had not had a major breakdown. This was attributed to the care which the members put into maintaining it. The secretary pointed out that all the moving parts of the windmill which require grease were always kept lubricated. Each member was said to contribute towards the buying of grease. Grease was bought in large quantities so that people never ran short. She said that sometimes they bought as much as 5 litres of grease which could last up to 2 years. The windmill was also said to be serviced every 4 – 6 months by the company that installed it. When a small problem develops a local mechanic, who is also a member of the nutritional garden, carries out the necessary repairs.

Some members attributed the good state of the windmill to the dedication of users who always made their contributions for its maintenance. However, they also attributed this to the nature of the technology itself. An observation which the water users made was that unlike boreholes which cannot 'operate themselves,' the windmill pumps water throughout the night into a tank even when there is no one at the garden. As a result users were almost always assured of water in the morning when they come to irrigate their vegetables. This eliminated scrambles for water.

The nutritional garden was said to be run along strict lines and rule breakers were said to be easy to deal with as there was a constitution which clearly spelt out disciplinary measures under different circumstances. Members are not allowed to join other irrigation schemes. It was felt that this reduced the chances of other households in the ward from getting irrigated plots. The garden was said to belong to the church. Every month members are obliged to pay a 'tax' to the church. The amount varied depending on the harvest. Most of the members were said to be have no problems with the tax. Half of the money which every member pays is ploughed back into the garden, for example to buy insecticides. The church uses the other half.

Case 3: Mpompini wetland: where diverse users converge

Wetlands are a crucial water source as they are dependable even during times of low rainfall or drought. Moreover, because of the proximity of water to the surface, water can be obtained without the need for storage facilities or water lifting devices. This means that there are low capital requirements, which is characteristic of informal irrigation. This is the case with the Mpompini wetland.

The wetland that is found in Ward 1 is an important source of water for households in Masiyepambili and Siyephambili villages. However, in times of severe water shortage, water users from as far as Thandanani village (which is in the same ward) also come to fetch water. Apart from being an important source of domestic water for the villagers, the wetland also supports a nutritional garden, Sidingulwazi Nutritional Garden, which covers an area of about 1.4 ha. The garden has about 117 members. There are five shallow wells on the wetland, three of which are used for drinking water while the other two are used for irrigation.

The wetland is used by three distinct water user groups who have competing claims to the wetland. The first user group is made up of domestic water users from Masiyepambili Village who base their entitlement on the fact that they live in the village in which the wetland is located. The wetland is regarded as communal property. It is on that basis that everyone in the village claims access to the wetland. The second group of water users is led by the committee members of the Sidingulwazi Nutritional Garden, and is made up of mostly women who have vegetable beds in the garden. Water users in this category are from Masiyepambili and Siyephambili Villages. They base their claim on the fact that that they were the first to use the wetland productively as an organised group with a committee in place. This committee has become the *de facto* authority at the wetland.

The third group of users is that of domestic water users from Siyephambili Village who base their claim on cultural morality. They are allowed to fetch water on the basis that water cannot be denied anyone. This perhaps is in recognition of the fact that the terrain in the village is not suitable for sinking boreholes.[13] The location of the village, on the

---

[13]The village has the least number of boreholes in the ward

edge of the ward, also means that households in the village are generally far from sources of water in other villages within the same ward.

Access rights generally overlap. Geographical residence is a factor in this - some of the members of the Sidingulwazi Nutritional Garden live in Masiyepambili and some in Siyephambili Village. Both groups of water users, those in the nutritional garden and those who are not, claim that Mpompini 'ngeyethu' (it is ours) but base their claims on different arguments. Members of the nutritional garden feel that since they use more water than domestic users they therefore have more rights to the water. The same members also have the only clearly recognizable institution in the wetland in the form of the committee. Members of the committee discuss access to the wetland on the basis of the nutritional garden. For example, when issues of who can be denied access to the wetland are mentioned, the most common reference is that if someone violates rules of the nutritional garden. They point out that since there is a committee and there are rules, those who violate these rules can be expelled from the garden. Thus, at least theoretically, it is possible for one to lose their use rights. However, because of the cultural moral values, the expelled member can maintain access rights to the wetland and still be able to withdraw water from the wetland for purposes other than gardening.

Rules around the use of wetlands are mainly to do with preserving the ecological integrity of the system and also to maintain certain standards of hygiene. Livestock owners are not allowed to graze or water their animals on the wetland. This is meant to avoid compaction of the wetland. However, livestock owners can water their animals just outside the perimeter of the fence around the wetland. People are not allowed to fetch water from the wells using metal containers which have been used on fire. This is to ensure that the wells are not contaminated with soot. Water users are also not allowed to wash their laundry within a specified distance from the wetland.

There are claims that there are no limits set to the amount of water one could fetch from the wetland. This was in recognition of the fact that some water users travelled long distances to the wetland to fetch water. However, those who brought large containers, or too many containers, had to give other users a chance and not fetch all the water they needed at once. Other rules apply more specifically to the members of the nutritional garden. For example, members are supposed to keep the gate closed to keep out livestock. As the dry winter season progresses sometimes members are obliged to reduce the number of beds they water. This is meant to ensure that the available water is shared fairly among the water users.

There is evidence of co-operation between users from Masiyepambili Village and Siyephambili Village. One example is obedience to the rules of access to water. Households from Masiyepambili Village, some of whom live within 200 m of the Mpompini, accommodate households from Siyephambili Village by giving them a higher priority when it comes to fetching water. They are also allowed to fetch water in large quantities. An indicator of how cooperation among water users enhances management of resources is that the fence remains in place and has not been stolen. Users only have to repair the fence periodically, which mostly requires labour input, not cash.

It is common to find water users from Masiyepambili Village fetching water in quantities as small as 5 litres, just to meet immediate domestic needs such as cooking. This reduces pressure for water at the waterpoint, which in turn enables water users from distant villages to fetch more water. Water users from Siyephambili Village, who walk as long as an hour and a half to get to the Mpompini, tend to fetch water to last them several days hence they bring more containers. In some cases households borrow scotch carts from their neighbours so that they can go and fetch water. Water users from Siyephambili Village use the water from the wetland strictly for drinking and cooking.[14] Unlike other water infrastructure such as boreholes, which demand that periodically water users make financial contributions for its repair and maintenance, the wetland lays no such demand. The most costly investment made on the wetland is the fence which was donated by the Department of Natural Resources[15].

Intensity of interactions among water users and between the water users and members of the committee varies with the season. In the dry season the wetland is at the centre of activities for households with vegetable beds in the garden. Since at that time of the year households are not busy with crops in the fields they spend most of the time in the wetland. Because shallow wells dry up in the dry season, this 'funnels' domestic water users to the wetland. This tends to increase contact between the different groups of water users. However, this was not observed to increase conflict among water users.

In the dry season there are also issues around livestock, which are not herded (they are allowed to roam since there will be no crops in the fields). They are attracted to the wetland because of availability of water and better pasture than can be found in the communal grazing areas. Keeping livestock at bay is one of the main tasks of water users. This entails maintaining the fence around the wetland. Because livestock sometimes break the strands of barbed wire, repairing the fence can be done as often as twice a month especially in the dry season. The tasks are mainly done by the members of the nutritional garden because they stand to lose more if nothing is done. Water users from neighbouring villages also participate in the deepening of one of the wells at the wetland which is prone to collapsing because it is not lined.

4.5 Summary of local water management characteristics in ward 1

Before presenting the discussion and conclusions of this chapter first a summary of the main characteristics of local water resources management in Ward 1 is given (Table 4.2). It is clear that there were more problems in case 1 (on the Gobalidanke borehole) than in cases 2 and 3 (the Smart cooperative garden and the Mpompini Wetland respectively). This could be attributed to the more severe water shortage that was experienced at the

---

[14] This strict use of water was in the narratives of the water users, and was confirmed in an extreme case in which the researcher discovered that one household routinely bathed toddlers in the household with 'recycled' water.

[15] Now known as the Environmental Management Agency

borehole than at the other waterpoints. There were, however, some similarities across the three cases. Cultural considerations were common in framing water access although other factors also came into the picture. This tended to be affected by the role played by outsiders. For example, in case 1 the involvement of an NGO tended to conflate issues around not just access to water but also the operational and maintenance arrangements. The evidence shows that the definition of 'local' varied across the three cases and were informed by factors to do with physical availability of water, how the water was abstracted and history of the water source. The definition of local affected contestations over access to water and, as can be expected, rules observance and conflict resolution arrangements. In general it can be concluded that local water management was affected by factors which could be social, physical and technological. As I will expand in the next section, there were no simple relationships between the factors.

Table 4.2 Overview of local water management characteristics in Ward 1

Parameter	Case 1	Case 2	Case 3
Definition of local	Ward based	Village based	Village based
Water availability	Poor	High	High
Competition for water	High	Low	Low
Operation and maintenance requirements	High	Medium	Low
Normative basis of accessing water	Cultural-based-but diminished by social differences	Cultural based-but enhanced by religious beliefs	Cultural-based but enhanced by environmental conservation
Origin of rules	Some rules introduced by project implementers, some by Environmental Health Technician, and others put in place by local water users	Rules derived from religious beliefs, and others put in place by management committee	Rules consistent with environmental protection laws, others based on traditional beliefs.
Role of outsiders	High	Medium	Low
Rules observance	Low	High	High
Level of co-operation among users	Low	High	High
Conflict resolution mechanisms	Waterpoint committee-but dysfunctional	Management committee-functional, church structures	Committee-functional, traditional leadership
Co-operation and conflict synergy	Poor	High	High
Institutional coordination	Low	High	High

4.6 Discussion and conclusions

The main objective of the chapter was to analyse practices in water resources management at the local level. However, given that the entry point for this chapter was the IWRM-based water sector reforms made in Zimbabwe in 1998, this section opens by relating findings of the chapter to IWRM.

A critical finding is that water resources management at the local level takes place without the influence of IWRM based catchment and subcatchment councils. This is rather surprising, especially considering the length of time which has passed since IWRM based water law and policy were promulgated in Zimbabwe. Among the changes brought by the new legislation was the partitioning of the country into seven catchments. Each catchment has a catchment council, below which are several subcatchment councils. Catchment and subcatchment councils were formed, among other reasons, to make water resources management participatory and to decentralise decision making in water management. Manzungu (2004) suggested that catchments are too large for catchment and subcatchment councils to be able to reach the local level and be effective. It is therefore surprising that close to a decade after Manzungu's study this problem has not been attended to. The fact that catchment and subcatchment councils are missing at the local level explains the absence of 'formal' IWRM at the local level and whypractices in water resources management at that level are (still) largely shaped by processes occurring outside of the IWRM paradigm. Among the major challenges facing Ward 1, Insiza District, are those relating to access to water for both domestic and productive uses, repair and maintenance of water infrastructure, and those concerning rule enforcement at waterpoints and conflicts among water users. One would expect catchment and subcatchment councils to spearhead the finding of solutions to these challenges. However, as was shown in the chapter, even water infrastructure development is being championed by state and non-state actors outside the catchment and subcatchment councils. Moreover, conflicts among water users are mediated by traditional/local institutions (such as village headmen) and local waterpoint committees. One would have expected catchment and subcatchment councils to play a lead role in such matters. One can hasten to add that waterpoint committees were the main institutional innovation in water resources management in the aftermath of independence (see Cleaver, 1990). As far as institutional structures in water resources management at the local level are concerned, it is notable that non-state actors who have provided water infrastructure to rural areas have to a large extent replicated waterpoint committees in one form or another. Form and function of the institutions have been retained. Thus, although the advent of IWRM was perceived as ushering in a new era in water resources management, the reality is that at the local level water resources management is continuing in more or less the same manner as prior to the arrival of IWRM. If, as observed in the chapter, IWRM-based institutions do not yet have presence at the local level, the question that arises then is, what chances are there that the challenges facing water users can be addressed within the foreseeable future through IWRM?

With respect to access to water resources at the local level, it was found that such access was guaranteed to everyone. This is because of the cultural framework within which water resources are managed. This was found to be true even at waterpoints where ownership and user rights were contested, such as at Gobalidanke. Although ownership of the waterpoint is contested, water users from across village and ward boundaries had access to it. Notably, it was also found that even at waterpoints where specific water user groups claimed ownership, such as at Mpompini, access to water was not denied to 'outsiders.' These findings are consistent with the observations by Nemarundwe (2003) and Derman et al. (2005) made in central and southern parts of Zimbabwe. They found similar cultural frameworks to be operational in different parts of the country. However, no matter how access to water was gained, users still had certain obligations which they had to fulfil. Such obligations included those of contributing towards the repair and maintenance of water infrastructure. It was found that at some waterpoints water user cooperation towards repair and maintenance of infrastructure was very low, while at other waterpoints this was much higher. This suggested that access to water as a factor on its own is not adequate to explain practices in water resources management.

As mentioned in earlier chapters, IWRM can be considered as one of several interventions which have characterised the water sector. At Gobalidanke waterpoint alone, for example, the fitting of the handpump to the shallow well represents one of the interventions in Ward 1's 'water sector'. Such interventions contribute towards shaping practices in water resources management in particular ways. Sithole (2001) and Kujinga and Jonker (2006) have shown that socio-historical factors play an important role in the management of water resources. At Gobalidanke waterpoint, for example, water user relations and consequent practices were influenced by a history of poor relations among the water users from the different communities. Interventions by non-state actors, specifically the fitting of the handpump, and the reclamation of the Gobalidanke gully, brought new dynamics to an already bad situation. The handpump meant water users had to contribute financially towards repair and maintenance of that particular piece of equipment. The fitting of the handpump also meant that users had to observe certain rules, such as not banging the handle of the pump on the ground when fetching water. However, rules were not observed. Water users generally did not cooperate over repair and maintenance of the handpump at the waterpoint. Other rules, such as not fetching too much water, not banging the pump on the ground, and even that of not doing laundry near the waterpoint, were openly broken. Such observations raise the question, how will IWRM as an intervention combine with other contextual factors in localities where it is being implemented, and with what consequences on practices in water resources management? The observation also suggests that it is imperative that before interventions are made it is important to understand contextual factors, in particular the latent socio-historical factors. These factors are often very localised but pivotal in determining the success or failure of interventions.

Apart from socio-historical factors, physical factors such as water availability were also found to influence practices in water resources management. At waterpoints where there were fears of water shortages the dynamics of water user interactions were different from waterpoints where no such fears existed. This was evident at Gobalidanke borehole where water users commonly ignored the rule on maximum water quantities drawn. However, it was also found that there appears to be a relationship between productive use of water and practices in water resources management. Arguably, all technology used to pump water, no matter how simple or complicated, requires repair and maintenance of some sort. This is true for handpumps and windmills, which are two of the technologies found in Ward 1. It was observed that cooperation among water users tended to be higher at waterpoints where water was used productively than at those waterpoints where it was used only to meet domestic needs. This can be explained by that productive uses can increase household income, which in turn increases the capacity and willingness of the household to contribute towards the repair and maintenance of the infrastructure. In the light of this, therefore, it can be argued that the tragedy of water resources management in the IWRM era has been too much concentration on institutional aspects of water resources management. This has been at the expense of issues such as attacking rural poverty, for example, through the provision of water for productive uses. Reducing poverty and increasing household income can potentially contribute towards improved water resources management as households become more able and willing to take on their responsibilities.

In conclusion one can say local water management is highly complex and dependent on a number of factors. Analysis of practices in water resources management at such a level shows that cultural factors, which emphasise empathy for other water users, provide the framework through which access to water is gained. This framework does not absolve water users of responsibilities, such as those of maintaining water infrastructure. However, for actors to fulfil their obligations they have to engage with each other, and that engagement is enhanced or constrained by socio-historical factors, and physical factors relating to infrastructure and water availability. Furthermore, whether actors cooperate with each other in managing water resources and infrastructure also appears to be influenced by the use to which water is put. Productive use of water increases the value of water which in turn facilitates cooperation (positive engagement) among the water users. Cooperation among water users facilitates effective maintenance of the water source which in turn positively influences the reliability of the water source.

Chapter 5

Livelihood strategies and environmental management in the Zhulube micro-catchment

5.1 Introduction

Since the beginning of the new millennium, environmental sustainability has been re-introduced to the development agenda. The Millennium Development Goals (MDGs) defined a specific goal for it (goal number 7), and formulated quantifiable targets: a) integrate the principles of sustainable development into country policies and programmes and reverse loss of environmental resources, b) reduce biodiversity loss, c) halve by 2015 the proportion of the population without sustainable access to safe drinking water and basic sanitation, and d) achieve significant improvement in the lives of the urban slum dwellers. It appears that the goal assumes a relationship between livelihoods and environmental management. This chapter explores this relationship in the Zhulube micro-catchment that falls in the Mzingwane catchment area (for a description of the study area see chapter 2).

This chapter analyses how local actors try to 'sustain' the environment and livelihoods in the context of a semi-arid micro-catchment and an adverse socio-economic environment. It is important to note that in this chapter, although the study area is described as a micro-catchment, delineated by hydrological boundaries, interventions led by non-state actors are generally based on administrative units. The focus is on the relationship between livelihood sustaining activities of residents and the physical resource base, commonly referred to as the environment. In this chapter I present an analytical framework that does not separate people from the 'environment. The chapter determines what types of resources are derived from the 'environment' and how these contribute towards livelihoods. The chapter also analyses to what extent local actors are conscious and take steps to protect the environment which is critical to their survival, and how this is shaped by outsiders, whether these be state or non-state actors. In other words, to what extent is environmental sustainability being pursued in the micro-catchment? The main question which the chapter tries to answer is: *how do local actors try to 'sustain' livelihoods and environmental management in an adverse socio-economic and natural environment?*

The analysis in this chapter is made against the background that in the first decade of the new millennium Zimbabwe has experienced a rise in poverty to unprecedented levels (UNDP, 2008). In this chapter poverty is discussed in terms of food insecurity, which in Zimbabwe is one of the clearest indicators of poverty. The 2009 Zimbabwe Vulnerability Assessment Committee (ZimVac) Rural Household Livelihoods Survey estimated that

1.4 million rural people in the country were food insecure[16]. More worrisome, poverty was now reported in areas where previously it was not prevalent. For example, a similar survey conducted in urban areas found that 33% of households did not have access to adequate food (Zimbabwe Vulnerability Assessment Committee, 2009). All the same poverty was reported to be more pronounced in semi-arid rural areas of Zimbabwe (Bird and Shepherd, 2003). This also applies to the Mzingwane catchment.

Since environmental sustainability is about the relationship between people and the physical resource base, the next section examines literature that addresses the relationship between poverty and environmental degradation. The section also discusses the analytical framework of the study, namely environmental security (Barnett, 2001). The context of the study is provided in the form of a historical account of how successive governments in Zimbabwe have tried to address environmental degradation. The empirical section that follows is made up of two cases that are situated in one and the same catchment area. The first case deals with gold panning, which in recent times has become a major source of income for a significant segment of Zimbabwe's population but is blamed for serious degradation of the physical environmental (Maponga and Ngorima, 2003). The case focuses on the actors that are involved, their livelihoods and the attendant environmental impacts. The second case analyses attempts at rehabilitating a gully in the same micro-catchment, showing how divergent interests have converged on a local environmental issue, and with what degree of success. The discussion and concluding section focuses on the relationship between livelihood security and environmental sustainability. The section shows that livelihood security is the most important factor determining how the poor relate to the physical environment. Securing livelihoods takes precedence over protecting the physical resource base. The section shows the importance of combining the theory of political ecology and the concept of environmental security in analysing environmental issues.

It is important to note that the two cases presented in this chapter to an extent illustrate the connections within the catchment, which are at the one level physical, and at the other level socio-economic. Within the catchment blue water resources are concentrated in the downstream part of the micro-catchment. The 'largest' small dam in the catchment, which is shown on Figure 5.1, is located at the downstream end of the micro-catchment, which is logical since the dam captures runoff generated in the upstream parts. There is also a wetland downstream of the catchment. The relative availability of water resources in the downstream part of the Zhulube micro-catchment explains the location of small gardens and the Zhulube irrigation scheme within the micro-catchment. The physical reality of more water availability in the downstream part of the micro-catchment is the reason why it is in the interest of the downstream communities to protect the upper part of the catchment, which is where gully formation and expansion processes are taking place. Social facilities, such as schools, clinics, main roads linking the micro-catchment

---

[16] Another dimension of poverty in Zimbabwe was revealed by the 2003 Poverty Assessment Study Survey (PASS II). The study showed that 72% of the country was living below the Total Consumption Poverty Line (UNDP, 2008). Although the findings of the PASS II study are not comparable to those of the Rural Household Livelihoods Survey which are used in this chapter, they nevertheless give an indication of the extent of poverty in the country through different sets of poverty indicators.

to other parts of the district, electricity lines, and shops and bars are located in the upper part of the catchment. Some of the gold panning taking place in the micro-catchment is located in the downstream part of the catchment.

5.2 The relationship between humans and the environment: some conceptual insights

Zimbabwe's Environmental Management Act of 2002 (Chapter 20:27) defines the environment as comprising of three elements, which are: a) the natural and man-made resources, physical resources, both biotic and abiotic, occurring in the lithosphere and atmosphere, water, soil minerals and living organisms, whether indigenous or exotic, and the interaction between them, b) ecosystems, habitats, spatial surroundings and their constituent parts, whether modified or constructed by people and communities, including urbanised areas, agricultural areas, rural landscapes and places of cultural significance, and finally, c) the economic, social, cultural or aesthetic conditions and qualities that contribute to the value of the natural systems and the man-made systems. The same Act defines environmental protection as the prevention of pollution or any activity which can lead to environmental degradation. While the above given definition of the environment appears to be all-encompassing, the Millennium Project observes that the causes and manifestations of environmental insecurity are so diverse that they cannot be captured in one definition, which is why it is important to focus on components of environmental security such as food security, water security and similar issues which affect mankind. This in a way calls for human-centred perspectives of environmental issues, which is not the same as ecological security where the biophysical resource is the point of reference.

A broader conceptualisation of the environment as presented above appears to be reflected in the manner in which watershed management is conceptualised by agents of development, especially non-state actors. Originally watershed management was conceived as the management of biophysical resources (soil, water and vegetation) in medium or large river valleys (Farrington *et al.*, 1999). The aim was to prevent or reduce soil erosion by reducing storm flows. This would contribute towards reduced siltation of reservoirs. This approach to environmental management has evolved, shifting from a focus on biophysical resources only to include livelihoods and institutions, and dropping in scale to operate at micro-catchment level (*ibid*). Farrington *et al.* also state that watershed management has now come to include enhancing the productivity of resources in ways that are ecologically and institutionally sustainable. Watershed management from this perspective is also about enhancing rural livelihoods. Such a view of watershed management is employed in particular by non-state actors.

In academia, the challenge of explaining the relationship between poverty and the environment is not new. The World Commission on Environment and Development (1987) came up with the concept of sustainable development, which, despite its wide criticism, has been influential in bringing environment into the development agenda. The Commission observed that there was an imbalance between population growth and the natural resource base, and that much of the global population growth was taking place in low income countries which also happened to be 'ecologically disadvantaged.'

However, the presumed straightforward Malthusian relationship between population and the environment has been challenged by other schools of thought. One such school of thought suggests that population pressure can actually force people to be more innovative. Innovation in turn can lead to increased productivity (Boserup, 1981). According to this view changes in production methods (technology) can actually result in less damage to the environment. This argument was supported by Tiffen *et al.,* (1994) whose study of Kenya's Machakos district showed that although population in the district increased fivefold between the 1930s and 1990s, soil erosion reversed in the same period. Taking a similar line of thought, Forsyth *et al.* (1998) cite examples from Northern Thailand, Papua New Guinea and Guinea and other countries where it was proven that population growth and poverty do not always result in environmental degradation. They argue that the poor are able to adopt protective mechanisms through collective action which can avoid environmental damage.

Whatever school of thought one subscribes to there is no doubt that that there are complex dynamics between poverty and the environment. If solutions to environmental degradation are to be found there is need for a better understanding of how poverty and the environment are dynamically linked (Nkonyo *et al.*, 2008). In this regard the focus on livelihoods is key. Moyo (2001) observes that the poverty and environment link has not been clearly exposed in the past because studies have mainly focused on calculable aspects of degradation, such as measuring quantities of soil erosion and the time frames in which deforestation occurs (e.g. Elwell, 1981). While such studies have shed light on human impact on the environment, they do not factor in the broader livelihood strategies of the people concerned. Such studies present man as the main cause of environmental degradation, and while this is to an extent true, there is need to, as the environmental definition presented earlier shows, consider that people are part of the environment, and that they can respond to environmental changes, and even anticipate such changes. One other problem is that such studies fail to show how environmental issues at the local level are, in a way, consequences of socio-economic processes at the national or even higher levels. In this regard political ecology makes an important contribution. Blaikie (1985) argues that explanations for environmental degradation at the local level, such as soil erosion, can be found in the link between land use patterns by local farmers and the international demand for agricultural commodities on the international market. Thus environmental degradation can be understood better when analysed within the political-economic framework. For this study it is important to analyse how the poor in a semi-arid environment have responded to socio-economic change and attempt to balance livelihood strategies and environmental sustainability.

The increasing importance of environmental security has been linked to the realisation that in the 21st century threats to peace and stability are more likely to come from environment-related issues as opposed to so-called traditional threats such as invasions by foreign armies (Pirages, 2011). This realisation has become even stronger in the post-Cold War era in which the danger that there will be war between the different ideological blocs has been reduced (Allenby, 2000). In fact it has become apparent that more people are dying of problems related to ecological crises rather than wars, for example (Pirages,

2011). Even in popular discourse it has come to be acknowledged that environmental issues pose a security risk. At the 13th Economic Forum held in 2005, for example, it was noted that environmental refugees, that is people failing to secure their livelihoods due to environment-related challenges, were on the increase (Myers, 2005). Within sub-Saharan Africa millions of people annually fail to secure their lives because of drought, soil erosion, desertification, deforestation and other environmental problems (*ibid*). These factors combine to induce poverty on the region's population. Thus there is need to consider seriously ecological security problems, such as those arising from the disruption of the equilibrium between environmental resources and population. From an environmental security perspective, it is necessary that policy reflects that the environment rather than foreign invasions is a more potent threat to peace and stability of nations. Proponents of environmental security also argue that policy should pay more attention to the human-environment relation rather than the traditional causes of war (Pirages, 2011). The basic point is that environmental issues are primarily issues of human welfare and security, thus poor management of environmental resources can lead to human insecurity. Environmental security as a concept therefore expands the definition of environment beyond the narrow understanding of environment as a physical phenomenon, and is thus able to factor in livelihoods of the people concerned.

5.3 Environmental change in Zimbabwe

Historical origins of environmental degradation in Zimbabwe

Looking into Zimbabwe's past can help in the understanding the environmental problems which the country currently faces. Resource use patterns, the environmental consequences of resource use, and the response of the colonial government in efforts to curb environmental degradation can be understood better when looked at as result of contestations, negotiations and alliances which the colonial government built with different actors (McGregor, 1995). This view is supported by Wilson (1995) who argues that local patterns of resource use and practices were deeply embedded in local politics and struggles with the government.

The colonisation of what is now Zimbabwe by the British South Africa Chartered Company was to a large extent influenced by mineral discoveries and speculation about a 'second Reef' (Hoadley *et al.*, 2002). Huge mineral deposits had been found in 1886 in the Witwatersrand area of South Africa and the white settlers had hopes that north of the Limpopo River pickings were equally rich. However, upon entering the country, the settlers discovered that mineral deposits were not as large as they had anticipated. Although the mineral deposits were disappointing, vast tracts of land ended up being cleared to support an emerging mining industry since the mines that were opened depended on timber for fuel and construction (McGregor, 1995). Furthermore, timber contractors preferred to cut down trees in areas set aside for Africans because by doing that they could escape from paying taxes (*ibid*). This contributed to processes of environmental degradation in what later would be known as Communal Areas[17]. In the

---

[17] See Chapter 4 for an explanation of Communal Areas.

areas set aside for Europeans trees were protected because the owners of land had title to it and could therefore keep out timber companies (*ibid*).

The failure to find minerals in the quantities that had been anticipated led to agriculture becoming the cornerstone of the economy. Agriculture supplied a growing urban population with foodstuffs and an expanding manufacturing industry with raw materials. Export of agricultural commodities also provided the country with much needed foreign currency. McGregor (1995) states that agricultural practices of white settlers were damaging the environment and threatened wetlands and river banks. More importantly, basing the economy on agriculture meant taking land away from the indigenous populations. The Rhodesian settler government pursued a policy of segregation. The policy involved two differently paced processes of modernisation along the lines of a dual economy that was both geographically and racially separated. This set in motion processes of accumulation by dispossession. The Land Apportionment Act (1930) racially divided the country by setting aside land for whites and for blacks. The objective of the colonial administration was to squeeze the indigenous population into reserves where they would form a class of smallholder farmers (Bolding, 2004). These smallholder farmers would practice a combination of intensive crop and livestock production. In the reserves agricultural production was not supposed to be detrimental to the natural resource base (*ibid*). This was contradictory since colonization had not only disrupted the agricultural system which the indigenous populations had developed over the years, but had also taken away the land resource on which agriculture was based. Practices which indigenous populations had used to conserve resources included land fallowing and seasonal grazing. These were no longer practical because of the changes in land ownership and size of landholdings which blacks had. Colonisation led directly to the bringing into the production system land which hitherto had been considered marginal. It is important to note that whites occupied the fertile areas while blacks were allocated marginal land in terms of soil fertility and rainfall.

Given the marginal resources in the areas set aside for the indigenous population, which was growing, the colonial administration considered an environmental disaster inevitable. These fears were bolstered by reports coming out of places such as the United States where southern and south-western states of Oklahoma, Arkansas, Kansas, Colorado and Texas among others, were said to be experiencing so much soil erosion that the food situation of the country was threatened (Pretty and Shah, 1997). News of the development of the Dust Bowl in the United States pointed towards the need to take action to prevent a similar situation developing in then Rhodesia. These developments led to the rise of conservationist concerns and the need to address environmental management issues. Officials became increasingly concerned with over-population, livestock increasing beyond the carrying capacity of the natural resource base and the imminent erosion menace in the reserves. At the time it was estimated that 16% of all arable land was degraded by sheet and gully erosion (*ibid*).

The colonial government responded to the impending environmental threats by putting in place legislation such as the Native Reserves Forest Produce Act (1929) and the Game and Fish Preservation Act (1929) (McGregor, 1995). The Native Reserves Forest Produce

Act (1929) banned tree cutting for any purpose other than to meet subsistence needs, protected species considered by the government to be valuable, and sought to regulate timber concessions. Discourse surrounding the enactment and implementation of such legislation placed blame on Africans who were claimed to misuse the environment *(ibid)*. This justified differences in the unequal restrictions which the legislation placed on resource use between blacks and whites. The government also decided to deal with environmental degradation through agricultural modernisation (Bolding, 2004). Thus the 1940s marked a shift in government policy in the demonstrator programme from livelihoods improvement to environmental conservationism *(ibid)*. The state took the stance that agricultural modernisation in the reserves had to be enforced. The demonstrator programme started as a method of improving agricultural productivity among African smallholder farmers by 'teaching through doing.' From the 1950s onwards smallholder farmers in the Reserves were forced to construct contour ridges as a means of controlling soil erosion.

As degradation worsened, the colonial administration decided to enforce the construction and maintenance of mechanical conservation works by law (Hagmann, 1996). Contour ridges were thought by the settler government to improve the mechanical protection of arable land from rill and gully erosion. The Native Land Husbandry Act (NLHA) and its forceful implementation can be seen as a massive attempt by the expanding technical bureaucracy to enforce the modernisation project. It was supposed to halt environmental degradation in the over-populated and overstocked reserves and put African agriculture on an environmentally, economically and agronomically sound footing (Bolding, 2004). The NLHA also sought to put a stop to labour migration between reserves and urban areas and to halt further settlement in the reserves by issuing saleable land and stock rights to a permanently limited number of African farmers.

However, government interventions had unintended consequences, as happened when there was an attempt to control the use of *dambos* (wetlands). *Dambos* are an important water resource which makes them vital in the smallholder farming system. In Zimbabwe dambos make up to a third of the land in the central watershed region of the country (McFarlane and Whitlow, 1990). In the period that followed colonization the government enforced a ban on dambo cultivation as fears of land degradation increased. Moreover, it became illegal to cultivate within 30m on both sides of a natural waterway (Hagmann, 1996). However, this led to the taking up of more upland areas for dryland cultivation and cattle grazing became more focused on dambos as pasture was converted to arable land. Resultant reduction in herbaceous cover increased runoff and reduced percolation. This consequently negatively affected baseflow which fed the wetlands. Increased run off also accelerated soil erosion.

Some of the conservation measures which have been enforced in Zimbabwe and also in other parts of the developing world are exotic, which in some cases has made it difficult for local farmers to adapt them to their own needs. Pretty and Shah (1997) state that graded and contour bunds, which are the hallmark of conservation works in smallholder farming areas, were developed in the United States of America for large scale farms. There they were constructed using heavy earth moving equipment, which was not available to local farmers. Pretty and Shah make the point that even in the United States maintenance of these conservation works was difficult because farmers neither had access to, nor control of this equipment. Such universal solutions did not appreciate the fact that solutions are environment specific in that in different environments there are local socio-ecological factors which can limit the transferability of solutions from one context to another (Hagmann, 1996). The layout of conservation works promoted in Zimbabwe, which consisted of contour ridges, drains, and waterways, was inadequate to control rill erosion effectively (*ibid*). Bad contour ridging was actually worse than no ridging at all (Pretty and Shah, 1997). In Zimbabwe farmers reportedly stated that ridges connected whole fields and drained into a single line. As a result during severe storms the ridges concentrated water into powerful and fast moving bodies that worsened erosion.

The above shows that in Rhodesia livelihoods featured in environmental management as a fringe factor. The concern of the colonial government was that indigenous populations should be able to feed themselves without destroying the environment. This dual objective would save the government from the burden of feeding the blacks and having to offer them more land.

Environmental degradation and poverty in the independent era

The historical factors outlined above to an extent account for the distribution of environmental degradation within Zimbabwe. A survey conducted by the Department of Natural Resources in the 1980s showed that 1.8 million hectares of the country were eroded (Whitlow, 1988). Of these, about 1.5 million hectares were in the communal areas where smallholder farming takes place. Only 270,000 ha of the degraded land were in General Lands, which were dominated by large scale commercial farming. Table 5.1 shows the percentage of each land class by tenure affected by different levels of erosion.

Table 5.1 Extent of erosion in different land classes

Erosion class	%Communal lands	%General lands	%Non-agric. lands
Negligible – Limited	60.0	95.2	98.8
Moderate	13.1	3.2	0.6
Severe	9.9	1.0	0.2
Very severe	17.0	0.5	0.3

Source: Whitlow (1988)

However, it is interesting to note that upon attaining independence in 1980 the Zimbabwean government embarked on a land redistribution programme which aimed at redressing the inequities in land ownership. Among the measures taken to safeguard against land degradation was to create the office of Resettlement Officer in the resettlement areas (Chigwenya, 2010). These officers were not only responsible for agricultural activities within the resettlement areas, but they also controlled resource use (*ibid*). Efforts of the Resettlement Officers were complemented by parastatals such as the Forestry Commission which helped the newly resettled farmers to establish woodlots in the resettlement areas. Such efforts, although they did not completely stop the degradation of the physical resource base, contributed towards improved resource management. Furthermore, when the government introduced the decentralisation policy which created Village Development Committees (VIDCOs) and Ward Development Committees (WADCOs), it emphasised that the committees also should have one person responsible for environmental issues. However, these initiatives, which had had a positive impact on the environment, were affected by the crisis which affected the country. From the late 1990s as the relations between Zimbabwe and the donor community soured, external funds which had supported the land reform dried up. The post of Resettlement Officer was abolished (Chigwenya, 2010). Resettled farmers were left without close monitoring and in some cases extended their fields onto grazing lands and other fragile environments. Smallholder farmers from neighbouring communal areas encroached into resettlement areas, cutting down trees and grazing their cattle. Such processes resulted in environmental degradation expanding into the resettlement areas which had hitherto been white farming areas and relatively safeguarded from degradation. In the 2000s the country experienced a severe economic crisis which left most government activities underfunded. Extension workers who used to have access to motorcycles and were thus able to monitor resource use in the rural areas were hit by a severe transport shortage and could no longer carry out this function.

On the institutional arena, among the changes which have taken in legislation affecting the environment include the enactment of the Water Act (1998), the ZINWA Act (1998), and the Environmental Management Act (2002). Notably, the Water Act (1998) created Catchment and Subcatchment Councils and conferred on them the responsibility of day-to-day management of water resources. Interestingly, although water resources management is the responsibility of the national water authority, the monitoring and enforcement of water pollution legislation is vested within the Environmental Management Agency (EMA), which is in charge of the environment.

Having traced the history of environmental degradation in the country the following sections move on to analyse gold panning and environmental protection in the Zhulube micro-catchment.

5.4 Gold panning in the Zhulube micro-catchment

To give context to gold panning activities in the micro-catchment, this section firstly presents the legal framework within which mining is carried out in the country, how this has changed over the years and how this in a way relates to environmental degradation in the micro-catchment.

5.4.1 Mining legislation in Zimbabwe

Gold panning is sometimes referred to as artisanal mining or small scale mining (Hilson, 2006). Whichever term is used, the activity referred to is a rudimentary form of mining (*ibid*). In a survey of eighteen countries across the world, the MMSD (2002) found that artisanal miners work with simple tools and equipment and mostly operate outside the legal regulatory framework. A common characteristic is that most of the miners in this category are poor and engage in the activity for subsistence (Hoadley *et al.*, 2002). Other characteristics of this type of activity were found to be the exploitation of marginal or small deposits, lack of capital, low rates of (mineral) recovery, poor access to markets and support services, significant impact on the environment, and low safety standards. Among the impacts of gold panning on the environment are mercury and cyanide pollution, direct dumping of tailings and effluent into rivers, river siltation, erosion damage and deforestation.

Although gold panning is currently considered to be an illegal activity, it is interesting to note that in the past the Zimbabwean government took steps to facilitate legal gold panning. But as will be shown below, that policy has not been consistently applied.

Legislation governing mining in Zimbabwe was to an extent shaped by the nature of mineral deposits in the country and the contribution of mining to the economy. Mineral deposits, especially gold, chrome, tantalite and precious and semi-precious gemstones found in the country often occur as relatively small deposits which are unsuitable for large scale commercial operations (Campbell and Pitfield, 1991). However, they are suitable for low-technology mining operations such as those of small-scale miners. In the case of gold small scale mines are considered to be ones which produce less than 15 kg annually (*ibid*).

In the 1990s the government gave gold panning a special dispensation which made it possible for individuals and cooperatives to enter the sector. Statutory Instrument 275/1991 (Mining Alluvial Gold) (Public Streams) empowered Rural District Councils (RDCs) to regulate alluvial gold mining (Zwane *et al.*, 2006). Under the system interested persons would apply for permits to carry out alluvial mining from the RDC. Prospective miners would have to be above the age of 18 and reside in the district in which the application is made. Licensed alluvial panners would be required to work some distance from the lowest point of the naturally defined banks of a stream, work without disturbance to the banks, not to dig trenches beyond a vertical height of 1.5m where there is no terracing and to rehabilitate the area they had been working on by backfilling at the end of their operations (*ibid*).

These regulations were critical because they specified areas where mining could be carried out and encouraged miners to restore areas to their original state through back-filling of pits (Maponga and Ngorima, 2003). The regulations theoretically enabled RDCs to control persons involved in mining since licenses could be revoked and previous records of the miners could be used to determine whether to allocate a license or not. This dispensation was a result of studies such as the one conducted by the Department of Mining Engineering at the University of Zimbabwe under a GTZ-funded project. The project proved that riverbank mining could be done in an economically and environmentally viable manner. It also demonstrated the benefits of rehabilitating mined areas in small-scale gold mining areas.

However, the dispensation which privileged gold mining and facilitated legalised gold panning came to an end roughly at the same time as the country's economic crisis worsened.

Under current legislation the process of becoming a registered small scale miner begins by obtaining a prospecting license, which gives one the legal permission to prospect for minerals in the country. There are two types of prospecting licenses, depending on the mineral(s) one is interested in. An ordinary prospecting license, which is for prospecting gold (up to 10 ha claims), or base minerals (up to 25 ha claim) costs US$50. A special prospecting license which is for base minerals only (up to 150 ha claims) costs US$100. One also needs a map which will accompany the application, which costs US$20. When one is issued with a prospecting license, one can prospect anywhere in the country. Upon finding the mineral(s), one then engages an Approved Prospector/Pegger to peg the mine. Such services cost about US$200. Once the claim has been pegged the prospective miner then applies for registration as a miner. Individual miners intending to mine gold pay US$50 for registration, or US$75 for base minerals. Once this is done a certificate of registration is issued. However, before mining operations can commence, the prospective miner has to carry out an Environmental Impact Assessment (EIA). The application form for an EIA, which is obtained from the Environmental Management Agency (EMA) costs US$210. After submitting the form, one needs to engage the services of a registered consultant to carry out the EIA, and such services range from US$1500-2000. Once the EIA has been done, it is submitted to EMA, which then charges US$1500 or 3% of the value of the minerals, whichever is higher. Once these processes are complete one can then operate legally as a miner.

The cost of becoming a registered miner (at least USD3500) is way above what most of the people involved in gold panning can afford. In addition the registration process is cumbersome. It is not far-fetched to suggest that the tightening of regulations governing mining has further driven gold panning underground, which in turn has led to the government further tightening its grip on the industry. The following section looks at how the government has tried to control gold mining in the country during recent years.

5.4.2 Attempts to control gold panning

In 2007 the Reserve Bank of Zimbabwe (RBZ) alleged that a lot of gold was 'leaking' from the country[18]. To curb these leakages, the RBZ established a unit which monitored gold production and marketing. The RBZ also set up gold buying centres in gold producing areas. All gold was supposed to be sold through these centres. Proceeds from the gold sales would be deposited into the seller's account by the RBZ. However, in practice it took long for the proceeds to be deposited into the accounts of the miners.[19] Worse still, the RBZ paid out part of the proceeds in local currency at the official exchange rate. The official exchange rate of the Zimbabwean currency to the United States Dollar at the time was a fraction of the rates offered at the black-market.

---

[18] The Monetary Policy Statement of 26 April 2007 by the Governor of the RBZ.

[19] The Monetary Policy Statement of July 2006 said that gold producers would be given a pre-export payment of 50% of the estimated value of gold delivered to the RBZ. The balance was to be paid within 21 days after the RBZ had exported the gold. Interestingly the statement of 2007 reduced the portion of foreign currency which the producers could keep from the 70% set in 2006 to 60%. The remainder was to be sold to the RBZ at the official exchange rate.

Consequently miners, both large and small scale, complained that their operations were becoming more and more constrained. These changes in procedures put in place by the RBZ were followed by '*Operation Chikorokoza Chapera*' (Shona for 'no more illegal gold panning') which aimed at curbing illegal gold mining and ensuring that all small-scale miners were registered. The operation had the effect of driving gold panners underground, making enforcement of sustainable mining practices difficult. These changes to the operations of small scale mining operators resulted in a sharp decrease in the number of those bringing their ore to the stamp mills for processing.

Operation Chikorokoza Chapera involved crackdowns by the police of what the government considered to be illegal mining operations (Spiegel, 2009). The government's stated rationale for the operation was to manage the environmental effects of mining and to control the diamond rush in the country's Manicaland Province. Miners who had not carried out Environmental Impact Assessments (EIAs) and did not have Environmental Management Plans (EMPs) had their operations closed. Spiegel argues that the focus of the government was clearly not on environmental degradation but on curbing smuggling. More than 30 000 miners were arrested throughout the country during the operation. In Insiza District panners claimed that the police operating in the area were quite vicious in their execution of the operation, using quad-bikes to penetrate the bush where the panners were operating. Initially this drove panning underground, and it became a 'nights only' activity. However, the panners claimed that this only lasted until the police began to accept bribes, and at that point the operation effectively died.

5.4.3 Gold panning as a livelihood strategy

Having looked at some of the changes to mining legislation in Zimbabwe, this section presents a case on gold panning in the Zhulube micro-catchment. Figure 5.1 shows the Zhulube micro-catchment and gold panning sites. Before that some background information is given.

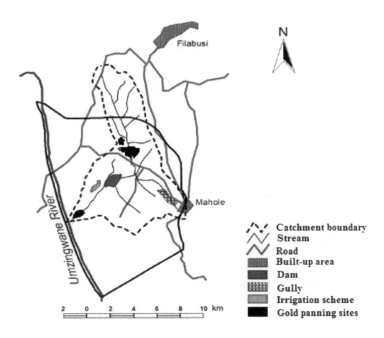

Figure 5.1 The Zhulube micro-catchment

In post-independent Zimbabwe the first boom in gold panning occurred in the early 1990s and was in direct response to the 1991/92 drought (Maponga and Ngorima, 2003). The second wave occurred in the late 1990s, and was related to the economic downturn in the country. The late 1990s were marked by the implementation of the Economic Structural Adjustment Programme (ESAP). The programme resulted in reduced government spending and a general economic decline characterised by massive retrenchments in the private sector. The third (current) wave of gold panning started in the mid-2000, and was associated with a combination of a series of droughts and a collapsing economy characterised by hyper-inflation among other things. These events had a national magnitude but it is at the household level where they impacted on income and food security and forced households to turn to gold panning.

In Ward 1 several factors were given as making gold panning a viable livelihood option. Firstly, it was said that money from gold panning comes 'fast.' Gold panners pointed out that although one could work for two or three days without getting anything, persistence was usually rewarded. The beauty of gold was said to be that you did not need to have a lot of it to get money. Buyers in the area were said to be willing to buy whatever gold one had on offer. The smallest transactable unit of gold was what was referred to as a 'point.' A point is equal in weight to a match stick, and on average it fetched about R100 or US$10. That amount of money was sufficient to buy about 5 kgs maize meal, cooking oil, salt, and a bar of soap. It could also buy about ten pints of beer.

Another factor which was given as making gold panning an attractive livelihood option was that one did not need much capital investment to start panning. Rudimentary tools, namely a digging implement and a small dish, a large sieve and a piece of cloth to trap gold during the separation process, were all that was required. In comparison crop cultivation was said to be more problematic. For example, it was said that if a household did not have oxen for draught power the household would have to wait until after those with livestock had finished cultivating. By that time the rains could have been gone. This dependence on the goodwill of others was considered inconvenient. Gold panning was also said to be an attractive option because the metal keeps its value. In a hyper-inflation environment panners stated that it was better to keep gold than the Zimbabwean currency which lost value every day.

In the ward there are stories of panners who have been extremely lucky, and have managed to buy livelihood assets. The best known case of a successful panner is that of Xolani. He is said to have found a particularly rich spot and in no time managed to buy a second hand car. But his is not an everyday case. More common are stories of panners who bought corrugated iron sheets to roof their houses and some who bought farm assets such as scotch carts, wheelbarrows, or livestock using proceeds from the activity. Such uses of money were associated with panners who were middle-aged, married and had children. Younger ones, especially the unmarried, claimed that apart from buying basic necessities, they also needed the money for fashionable clothes and entertainment. There are stories of panners who travel all the way to Bulawayo, spend the night in nightclubs and return the following morning after having spent all the money they had.

Gold panning supports the local economy. Apart from the stimulation of local businesses, gold panning also created demand for services which otherwise would have few or no customers. One such service was the need for transport. In winter, when the streams in the ward are dry, gold panners hire scotch-carts to ferry their ore from wherever they would be operating to small reservoirs in the ward where they wash their ore. As a result households that had scotch-carts benefited from gold panning. Some households claimed to be making as many as 6 trips to the dam every day[20]. Areas where panners operate attracted women who sold foodstuffs[21]. Irrigators also relied on gold panners to buy their produce. A belief among panners was that to get more money you had to spend what you had. Panners themselves said that there was no other activity which could earn them more money than gold panning.

Within the Zhulube micro-catchment gold panning has become a very important livelihood strategy for most households. At the peak of the 2007/08 drought, village head Nonto estimated that more than three-quarters of the households in the ward were engaged in gold panning. Panners that were interviewed claimed that every household

---

[20] Since the ore is very heavy, some, however, regarded this business as a cruelty to donkeys which pulled the carts.

[21] In the periods that followed rumours of someone finding a particularly rich vein, it was common to find women selling meals since the panners claimed they did not have time to waste preparing meals. However, when it became apparent that the finding was just an isolated lucky strike, the panners would revert back to buying only what they needed to prepare their own meals.

had at least one member engaged in gold panning. Even civil servants in the ward were said to be engaged in gold panning. It was common to come across boys of school going age panning for gold during school hours. Teachers at Tshazi Secondary School reported that some drop outs from the school were linked to gold panning[22]. Another claim which the panners made was that even traditional leaders were involved in the activity, though secretly.

To show how environmental issues and livelihood strategies inter-lock, three gold panners, Mr. Mbongeni Ndlovu *(a college drop-out turned gold panner)*, Mrs Madubeko *(widow, grandmother, care giver and gold panner)* and Mr. Ndabezile Sibanda *(young and care free gold panner)* are profiled.

Mr. Mbongeni Ndlovu is 26 years old. He is married and has two children, a six-year old daughter, and a one year old son. He stays with his family just a stone throw away from his father's homestead. He attended Tshazi Secondary School. He has seven Ordinary Level passes, including Mathematics, English, and Science. His passes enabled him to enrol for a course in Environmental Health at Masvingo Polytechnical College. His hope was that upon qualifying he would be employed by the government's Ministry of Health and Child Welfare as an Environmental Health Technician. However, as the economic situation in the country deteriorated, fees at the college were raised. With the fees pegged beyond his means, he had no choice but to return home. That is how he ended up in gold panning. Initially his idea was just to raise enough money to pay his fees and complete his studies. However, because of the hyper-inflation, it was not possible to work once-off and be without income for extended periods. He found that even if he saved some money and returned to college, the money would not be enough to last him a term (which is about three months long). To worsen the situation, even the money he left with his family was not enough to sustain them while he was away. What he found discouraging was that his prospective employer, the government, was no longer attractive. He felt that it did not make sense to spend all his energy raising money to pursue a course which at the end of the day would lead him to a lowly paying job. After considering these factors he dropped out of college and returned home. At the time of the research gold panning was now the most 'stable' source of income he had. As someone who had some tertiary education on environmental issues, he understands the impact of gold panning on the environment, but he justifies his activities on the need to look after his family.

I met **Mrs Madubeko** while she and two children were panning for gold along an isolated stretch of the Insiza River. Although in the ward gold panning is not an exclusive domain of males, this was the first time I had seen an elderly woman engaged in this activity. It turned out that she was a widow and the children she was with were her orphaned granddaughters, one was in Form 1 (14 years old), and the other in Grade 6 (13 years old). She was working on a 'dump' left behind by previous gold panners. Before

---

[22] A number of cases of ill-discipline among boys in secondary school was said to be a result of the money they were getting from gold panning. One narrative was that a headmaster once tried to counsel a boy on the importance of attaining secondary education but the boy instead told the headmaster to shut up since he could pay him and all his teachers.

the economic challenges of the 2000s, although the state's welfare system did not reach the elderly in the rural areas, children of school going age had their fees paid for through the state funded Basic Education Assistance Module (BEAM). However, as the economic meltdown worsened this safety net also gave way, leaving people like Mrs Madubeko having to meet the educational needs of their households. She said that gold panning was mainly to meet the educational requirements of the children, such as school fees and textbooks. She received food hand-outs from World Vision and that really helped her. However, there were some grocery items which World Vision did not provide but which are essential, and these she had to buy for herself. Since her grandchildren are girls, and still very young, the three of them cannot be like the other young and strong gold panners who dig deep into the ground. Instead they re-worked what other gold panners abandon. In the rainy season they look for alluvial gold in the streams. Alluvial gold is easier to mine because it is not very difficult to dig, and also because they work after it has rained when there will be water in the streams, so they do not have to hire scotch carts to ferry their ore to the small reservoirs in the ward. An indicator of how the hard times had made everyone 'streetwise' was that when asked about how she sells her gold, and her personal safety in the transactions, she said that 'everyone knows where to sell gold.'

Ndabezile is 21 years old. He is the fifth child in a family of seven, being the younger of two sons in his family. His elder brother migrated to South Africa in 2001 and never phones, writes, sends money or groceries back home. His parents separated and he lives with his mother and now considers himself to be her breadwinner. Although his mother has some cattle, he would rather they keep them for serious emergencies than sell them to meet every day needs. He has never been formally employed, something which he attributes to the events which were taking place in the economy when he finished secondary school (Ordinary Level). At least twice a week he does gold panning, working the whole day. He says gold panning is very hard work which can kill. Sometimes it takes two or more days just to get to the gold bearing rock. If one is unlucky they can spend time digging but fail to get any gold, and this is why it is important to dig in the vicinity of where others are also digging and have proven there is gold. The hard work also makes it important that he shares the labour. Sometimes he gets a 'point' of gold a day, if he is lucky he gets more than that. He says that he is into gold panning because as a man he cannot depend on money from his mother, money that he needs for personal things such as clothing, beer and for entertaining his girlfriends. He has never invested the money he gets from gold panning to buy something which can be considered an asset, such as livestock, or a scotch cart. There is no reason to explain why he has failed to buy such things. His view is that gold panning is a useful activity because it is a way of earning an income, and it has also reduced criminal activities in the village.

Common to all the panners discussed here is poverty. All gold panners profiled stated the need to earn money to meet some needs, such as basic household nutritional needs and educational and other needs for children. Perhaps the saddest case is that of Mrs Madubeko, who in her old age has had to turn to gold panning to fend for herself and her grandchildren. Notable is that although gold panners still in their youth are potentially employable because of the level of education attained, the economic situation in the country reduced prospects for formal employment. Although the state considers such unregistered mining activities to be illegal, young men in the ward consider gold panning to be a better livelihood option than other illegal activities such as stealing or illegal migration to South Africa.

At the surface gold panning appears to be highly disorganised, each panner looking after his own interests. The activity is also commonly associated with crime or violent behaviour. It is common to hear about fights between or among young male panners. When someone's chickens disappear from the coop gold panners are usually the first suspects. It is also common to hear mothers complain about how panners 'play' with their daughters. While these things indeed happen, relations among the gold panners themselves are close and based on virtues that characterise ordinary friendships in life. Take for instance Vusimuzi, Mthokozisi and Ndabezile. These three young men were friends in high school, and in the post-high school life faced unemployment together. Mthokozisi and Vusimuzi were the first to go into gold panning. Ndabezile joined them later. Initially they worked separately. However, the hard work of digging the pits brought them together. Now they share the labour, and they also share the proceeds. The nature of gold panning demands that they work together. For example, sometimes they dig a deep shaft from which they may tunnel as they follow the 'scent of money.' In such cases team work is essential. Getting the ore to the surface demands two or more people working together - one person goes down the shaft, loads the ore into a bucket and the person on the surface pulls the bucket up. If one would do this individually it would be much less efficient. Sharing labour also enables some to work while the others rest or prepare a meal. Once they have enough ore, they fetch water together, helping each other to pump water at the borehole. When the state launched Operation *Chikorokoza Chapera* to clamp down gold panning it was also important that there was always someone at the surface looking out for the police and to inform others of any impending danger. Trust is key in such relations, and most panners said they were sure that it is not possible for others to cheat and not be found out. In such a rural setting the belief is that things always come out because the gold buyers are known and the community is small.

5.4.4 Impacts of gold panning on the environment

The most visible impacts of gold panning are the scars it leaves on the landscape. Pits which are dug can be as deep as 5m, some of which are tunnelled from the 'main shaft.' Material removed from the pits is left lying all over, only to be washed into river channels by the rains. Pits which are left uncovered are a danger to humans and livestock. Panners dig up river beds and banks in search of alluvial gold. Soil removed from these pits is left to choke river channels, and because the soil would have been loosened, when the rivers flow sediments are washed into dams. There is real danger that the Zhulube Dam (the water source of Zhulube irrigation scheme) will be silted as material is deposited into the dam (Figure 5.2). In fact the old Zhulube Dam was silted up, and the role of gold panning in that cannot be discounted. A study by Tunhuma (2007) indicated the presence of mercury in the Tshazi River, which he attributed to processes of recovering gold from ores. Figure 5.2 shows damage to the environment while 5.3 shows a gold panner going down a shaft.

Figure 5.2: Environmental damage caused by gold panning

Figure 5.3: A gold panner going down a shaft

When asked about the damage to the environment, most of the panners interviewed acknowledged that they were aware of the impacts of their activities on the environment. However, they said they had no choice but to pan to earn a living. Some of the panners also pointed out that they were willing to use 'environmentally friendly' panning methods but they said the formal system worked against them. The panners were aware of an experimental project which a German organization carried out together with the University of Zimbabwe on gold panning technologies which are environmentally friendly. They expressed willingness to be part of the project but did not know whom to contact. They said they did not know what was needed for them to get permission to operate legally. A few were of the opinion that they had to get a permit of some sort from the Environmental Management Agency (EMA), but they did not know what was required for them to get the permit. Getting information from EMA offices at Filabusi was said to be nearly impossible because the offices were said to be almost always closed.

5.5 The gully reclamation project in the Zhulube micro-catchment

While some local actors were busy with gold panning, which severely impacted a part of the Zhulube micro-catchment, other actors were at the same time involved in a soil conservation project elsewhere within the same micro-catchment. This was the Gobalidanke gully reclamation project.

5.5.1 The Gobalidanke gully and its place in culture

The Gobalidanke gully is one of the most notable physical features in the Zhulube micro-catchment. It spans about a kilometre in length. It is on average 2 m deep, being 2.8 m at its deepest point. At its widest point it is 28 m wide, its average width being about 20 m. The gully, which has an east-west orientation, is located just about 400 m south-west of Mahole Business Centre and less than 100 m to the south-east of the Zhulube Clinic. It is located upstream of the Zhulube Dam (Figure 5.2).

Gobalidanke has a special place in the culture of the local people. Narratives of elders in the area relate the place where the gully now stands to the Njelele rain making traditions. Ranger (1999) states that rain making ceremonies have been part of the Ndebele cultural traditions since their arrival in the present day Zimbabwe in the 18th century. The elders in the area highlight that in the past Gobalidanke was one of the sites where *'hosanas'*[23] going to the Njelele would stop to drink water because the place never dried up. On their way from the shrine *hosanas* would again stop at the waterpoint and soon afterwards the rains would fall. The belief was that the rains would wash away the footprints of those who would have gone to the shrine[24]. It is said that the area began to dry up in the period after independence when the whites, in their efforts to bring clean water to the community, used cement on one of the wells at Gobalidanke to protect it from contamination. Use of cement reportedly angered the spirits and as a result the area began to dry up until it developed into a gully. However, others have linked the development of the gully to the lack of enforcement of the ban on livestock grazing. As the cattle population increased the area was overgrazed which resulted in increased soil erosion and eventually a gully formed. I have not been able to establish exactly when the gully began to form, but DNR records show that already in 1983 the gully was so well developed that initiatives were taken to stabilise it.

5.5.2 Rehabilitating the Gobalidanke gully

Over the years there have been several attempts to rehabilitate the gully by both state and non-state actors. In 1983, for example, the Department of Natural Resources (DNR), the precursor of the Environmental Management Agency, initiated a project to rehabilitate the gully. The project lasted only a few months because funds ran out. Later in the same year a local NGO, ORAP, brought in some funds to resuscitate the project. Interventions by both the DNR and the Insiza RDC focused on controlling land use in the area around the gully. A fence was erected around the gully to enforce a ban on livestock grazing in its vicinity. Gabions (stones packed in wire mesh) were also placed on the gully to trap sediments and to prevent the gully walls from collapsing. However, judging by the fact that the gully continues to grow, and that the fence around it has been stolen on a number of occasions, these interventions have not been successful. Community leaders blamed the failure of the projects on the limited participation of the local people. They point out that technical experts would simply come and *'do their thing'* without consulting the local people for ideas. This was said to be wrong since the place is sacred. This view was

---

[23] Cleaver (2000) refers to *hosanas* as spirit mediums, but the researcher's investigations suggest that it is difficult to place the *hosanas* in the same category as spirit mediums since they act as emissaries who represent chieftainships at the Njelele rainmaking ceremony.

[24] The rain making ceremony can be considered to be a tradition in transition. Water users point out that *hosanas* going to the Njelele no longer follow traditions, such as walking and not using public transport to the shrine. Water users also see the *hosannas* as now being interested in money only. In one instance in 2008 the *hosannas* said that the 'Rock' (at the shrine) had told them the money which they had taken to shrine was too little. At the time Zimbabwe's economic problems were quite pronounced, and this led some members of the community to suggest that the demand for more money was a message from the 'Rock' but was related to the state of the economy and survival strategies of the *hosannas*.

also shared by the Field Officer of World Vision. About the failure of the efforts of the Department of Natural Resources he said:

> Efforts by the government failed because of lack of community education on the importance of reclamation. The level of participation by the community was very low, the project did not really belong to the people. The government did not consult the people on how to reclaim the gully. *(Filabusi, 13/06/2007)*

5.5.3 The food for work project

The most recent intervention in 2006 on the Gobalidanke gully was led by World Vision. Rehabilitating the gully involved mainly filling it with stones fetched from a nearby mountain. The idea behind filling the gully with stones was that this would stabilise it by breaking the flow of water and trapping sediments. Vegetation was also planted on its sides to stabilise the soil. It was said that these techniques were suggested by the Environmental Management Agency together with the agricultural extension agency AREX. A Field Officer with World Vision stated that the rehabilitation of the gully was done using stones to demonstrate to the community that they could protect the watershed using locally available resources, which are freely available in the environment.

Under the project, every week each of the six villages in the ward worked for four hours on the gully on its allocated day. The time was calculated to ensure that households would not spend the whole day working on the gully, but would have time to tend to their fields. Almost without exception participating households were represented by a female member of the household. The everyday supervision of work on the gully was delegated to traditional leaders, and members of the Village Management Board[25].

Initially the project was meant to be voluntary. However, it was soon noticed by World Vision staff that households were not coming to work on the gully. The reason which was given for this was that households were spending time looking for food. The NGO then decided to incorporate the reclamation project into their 'assets for food' programme and households working on the gully began to receive food hand-outs from World Vision.

To qualify to work on the project one had to be:
- able bodied;
- between the ages of 18-60 years old;
- vulnerable (defined as):
  o food insecure
  o no household assets such as cattle, plough, scotch cart
  o high ratio of dependants versus assets
- live within 5 km from the gully

---

[25] According to a Field Officer with World Vision, the NGO created Village Management Boards as a way of reviving Village Development Committees in areas where these had collapsed. These committees were formed for the purpose of implementing the NGO's projects, but were not meant to be in conflict with the conventional village structures. However, Village Management Boards were disbanded after the NGO was accused of creating parallel structures in the villages.

Although a World Vision Report states the above as the criteria for working on the project, visits to the project between 2006 and 2007 showed that in practice the criteria were not strictly adhered to. For example, irrigators from the Zhulube Irrigation Scheme, who are considered by other members of the community to be better off than rainfed farmers, also worked on the rehabilitation project. Male irrigators stated that their wives were the ones who went to work on the gully while they themselves tended crops on days when their respective villages were supposed to work on the gully. Targeting rainfed farmers had been meant to discriminate in favour of households considered to be most vulnerable to the effects of weather, and therefore most likely to be food insecure. Verifying the levels of vulnerability of households was said to be difficult. It was said that households could present a profile that suited whatever selection criteria an NGO chose to use. The only conditions which appeared to have been adhered to were that one had to live within 5 km from the gully and had to be able bodied. The latter condition was self-enforcing since work at the gully was physically demanding. The former condition was also easily enforceable since households beyond the 5 km radius generally tended to fall outside Ward 1. The project was meant to benefit Ward 1 and households in that ward[26]. People in Ward 1 stated that if 'outsiders' joined the project they would take places which could otherwise have gone to households from their ward. As a result, households from some villages in Ward 2, which is located upstream of the gully and are less than a kilometre from Gobalidanke, were not allowed to join the project. A total of 758 households were selected to work on the project, and these were from all the 6 villages in the ward. There were households from upstream and downstream of the micro-catchment. Figure 5.4 shows women filling part of the Gobalidanke gully with stones while Figure 5.5 shows a procession of women bringing stones to the gully.

---

[26] A similar point was made in Chapter 4. NGOs use the approach of working in wards, and in some cases together with the RDC select wards in which to operate in. The choice of wards is based on a number of factors, such as the nature and aim of the project and the perceived level of poverty in the ward. The rationale for this approach is that projects can go to those wards with the most need, and where projects will have the most impact. It also avoids cases whereby one ward can have a lot of donor organisations working in it while other wards are not receiving aid. However, NGOs tend to focus on the same wards they have been working in. While this may sound reasonable, in practice it can separate two adjacent wards which share the same environmental resources, have social and cultural ties, and face the same challenges.

Figure 5.4 Women fill a part of the Gobalidanke gulley with stones

Figure 5.5 Women form a procession as they bring stones from a nearby mountain to fill in the Gobalidanke gully

5.5.4 Impact of the project

World Vision outlined the objectives of the project as being to stop the expansion of the gully, to prevent the siltation of the downstream Zhulube Dam, and to protect infrastructure threatened by the expansion of the gully. However, to the community an important benefit of the project was that it met their food requirements. At the time of the project the country was experiencing an economic meltdown and a drought and households in the ward were in urgent need of food relief. Although the government was indeed trying to distribute drought relief to households in the areas affected by drought there were several constraints to getting the food to households. Fuel shortages contributed to the delays. In addition the government was distributing the maize through a parastatal, the Grain Marketing Board (GMB), which is based at the district level. Communities were obliged to find their own means of transporting the maize from the district headquarters to their villages. This entailed collecting contributions from individual households and then hiring a lorry and buying diesel so that the maize could be transported to the villages. It was therefore common for households to go for several months before they could get food aid[27].

Households that worked on the project were entitled to: 3 kg of cereals, 0.5 kg bulgur wheat, and 0.185 litres of vegetable oil per day worked. In practice households received 75 kg cereals[28], 10 kg pulses, 4 litres of vegetable oil per month. These rations had a huge impact on food availability at the household level. In fact one village head commented

---

[27] The initial response of the government to the droughts of the 2000s was to ban the movement of maize from one place to another. This was based on the perception that food shortages in the country were being caused by people who were exporting maize purportedly to increase the suffering of the people. As the food shortages worsened the government lifted the ban on the importation of maize by individuals and millers.

[28] This was later reduced to 60 kgs when the stocks of food and the funding for the project ran out.

that "*money that used to be devoted to the purchasing of food is now being used for other purposes, such as paying school fees for children, buying livestock such as goats and chickens.*" The village head also pointed out that because of the Gobalidanke project no one in the ward was facing starvation. A social indicator of food security which the project brought was said to be that at the time of the project it was not a problem for large households which ran out of food to be provided for by other households in the community. Smaller households were said to have enough food to spare, in some cases even selling food items such as cooking oil. The project also created an opportunity for households that had been left out of the Zhulube Irrigation Scheme to reap some benefits from World Vision. As mentioned above, the Zhulube Irrigation Scheme, established by the NGO, only has 42 households. The Gobalidanke project accommodated hundreds of households, and the community saw this as a way of spreading benefits from the NGO to everyone in the community.

Evidence of the support the project garnered, which also indicated the impact on livelihoods, was in the numbers of households that participated. During the period when World Vision was running the project, there was always a hive of activity at the gully, with line after line of women streaming up and down the mountain balancing rocks on their heads, or carrying them in their hands. Each day of the week women from a participating village would arrive around 9 a.m., and they would work steadily until mid-day. Although all supervision of the work was done by community leaders, and none of the NGO staff would be present, work was done faithfully without serious cases of repeated absenteeism being reported.

The next section profiles one participant in the Gobalidanke Rehabilitation project to illustrate the circumstances in which some of the participants were in, and how the project helped them.

5.5.5 Profile of a household benefitting from the rehabilitation of the Gobalidanke

To understand how the gully reclamation project became intertwined with the livelihoods of water users in Ward 1, the case of Mrs Mabheka is informative. Mrs Mabheka is an elderly woman, probably in her late 60s who grew up in Filabusi, having moved to the area in 1969 upon getting married. In 1989 her family moved to their present home in Thandanani Village. Among other things, she is part of the community leadership as a member of the Village Management Board and also as a faith healer affiliated with the Twelve Apostles Church. Her family used to live at Epoch Mine where her husband was employed. A mining accident left him with a severe back problem and he had to resign as he was no longer strong enough to do heavy work. The family moved out of the mine house and came back to their rural home, but the husband is not strong enough to work in the fields.

Mrs Mabheka had 11 children, the youngest of which is a daughter still attending primary school. Four of her children have since passed away. She does not hide the fact that some of them probably had HIV/AIDS. "*I think they had this disease which came,*" so she says. Two of her three remaining sons are seriously ill. They do not live with her, so occasionally she visits them and sends them whatever she has. She also has to take them to faith healers as she searches for a cure. Because of the tragedy that befell her family, she is the one who looks after six of the orphans left behind by her deceased children. Another of her grandchildren lives with her because the child's parents in town cannot afford the costs of sending her to school in town. Two of her grandchildren have had to be pulled from school because she could not afford to pay fees for them. She says that sometimes her homestead resembles a crèche because there are always children around. At the time of the research she was also looking after her sick sister. Food is the major problems her household faces. A major concern she has is that '*food does not stay here, it comes and goes.*'

The household has a field in which they grow maize, but because of the low rains they hardly ever harvest anything. Even when the rains are good, they cannot afford to buy seed and fertiliser. 'Those things require money', she says. As an enterprising woman she tried buying and selling vegetables, buying them at Filabusi or Esigodini and selling them at Mahole. However, since the bus fare is always going up the profit from the venture is now too small, and if the vegetables are not sold within a few days she incurs a loss. She sees the Gobalidanke project as a salvation. Now at least the household gets food for the grandchildren. Though the food hand-outs from the project sometimes come late, the household never really runs out of food. In fact she says that since the project began she has been able to provide her household with meals every day, not just once but twice a day. The household gets from the project much more than they can get from their fields. That is why she is so grateful for the project, not only for her own household, but even for child-headed households in the village. Mrs Mabheka's life represents one of the sad stories of the effects of the decline of the country, the impact of the harsh environment, and the ravages of disease on family life.

5.5.6 Sustainability of the Gobalidanke project?

A World Vision report (The Profiles of Lifetime Details: The Zhulube Irrigation Scheme, Insiza ADP, Matabeleland South Province, no date) claimed that its approach to the project was sustainable because the project involved the participation of the community. The report anticipated that the community would carry on with the project when the time for World Vision to pull out of the project came. This assumption was based on that the community could see the impacts of their efforts. However, in 2007 the funds allocated to the project were exhausted and the food rations which households had been receiving for working on the gully stopped. In the two years that followed work on the gully completely stopped. Not even a single person was observed working on the gully. At the sites where the gully was filled with stones and vegetation planted the gully appears to have been stabilised. However, at its head expansion is still going on, and further downstream where the gully was not filled with stones the gully appears to be still expanding and its sides collapsing. Interestingly the NGO went on to embark on a project that concerned the de-silting of small dams in the same ward. The project also handed out food rations to participants.

5.6 Discussion and conclusions

The relationship between human beings and the environment defies simplistic analysis, as illustrated by the two cases that were presented in the chapter, one on an environmentally damaging activity (gold panning) and the other one on environmental protection (the rehabilitation of a gully). The fact that these cases were observed within the same micro-catchment, and involved actors coming from the same households, is evidence of the complexity of that relationship. The chapter based its analysis of livelihoods-environment relationship on household livelihood strategies (Moyo, 2001), the argument being that efforts at obtaining food are central to household activities. It must be highlighted that in this chapter analysis was limited to the efforts made by households to meet short-term needs, which contrasts the idea of food security as defined by the Food and Agricultural Organisation (FAO) (1996). FAO defines food security as existing when '...*all people, at all times, have physical and economic access to sufficient, safe and nutritious food to meet their dietary needs and food preferences for an active and healthy life.*' The standards set by FAO in its definition clearly and emphatically show that households in the study area cannot be said to have achieved food security.

Empirical material presented in the chapter suggests that local actors, faced with an adverse socio-economic and biophysical environment, frame the environmental sustainability discourse strategically so as to serve their livelihoods agenda. The fact that the same actors, on the one hand ally themselves with non-state actors and engage in environmental management initiatives, but on the other hand engage in actions that completely oppose sound management of the environment seems to point to opportunistic behaviour. Local actors therefore can be seen as exploring and exploiting the institutional layers and the biophysical layers to try and put food on the table. Thus the actions of the local actors have little to do with environmental management, but everything to do with survival strategies under circumstances of extreme stress. This finding confirms the views of Vishnudas *et al* (2012), namely that for physical improvements to the watershed to be sustainable they must be tied to socio-economic benefits. They suggest that one way of achieving this can be through a payment for environmental services system. This can be achieved, for example, if downstream and wealthier people offer economic incentives to upstream people to preserve the resource base (*ibid*)

Although the chapter focused on the responses of the poor to an adverse socio-economic environment in the 2000s, it is important to highlight that to an extent the trend appears to show that policy prescriptions trigger unsustainable activities of the poor. In Zimbabwe, in the early days of colonialism, land degradation was attributed to poor agricultural practices of the Africans (Bolding, 2004) when in fact the major cause of soil erosion was the land alienation policy that the colonial government implemented. Reduced landholding size among the indigenous people compromised food security at the household level. This resulted in environmental problems at the local level since Africans could no longer continue practices that protected the physical environment from degradation. To be able to grow food crops to meet their needs, local actors were forced to bring into the agricultural production system all the land they could access, even if that meant utilising marginal land or ecologically sensitive areas such as wetlands. Use of ecologically sensitive areas resulted in degradation of the physical environment. The fact that marginal physical environments are brought into the production system for livelihoods purposes and the consequent degradation of the physical environment supports the views of Barnett (2001) that environmental issues cannot be separated from household welfare. However, one observes the different responses the degradation of the physical environment has triggered from different actors. The colonial government's response to environmental degradation was to put in place policy mechanisms which further restricted the use of environmental resources for livelihood purposes. Such policies included those that banned the use of wetlands and stream bank cultivation (McGregor, 1995). To this one can also add those policies that sought to coerce local actors to put in place structures which were thought to reduce soil erosion, such as contour ridges (Hagmann, 1996).

In the 1990s the country's mining policy facilitated the entry of small-scale miners, who included gold panners, into the mining sector. An important policy objective was employment creation since small scale mining does not require much capital investment (see earlier definition) and is labour intensive. Another objective of the policy was to create an environment conducive for small-scale miners to exploit mineral resources in a manner which did not result in significant damage to the physical environment. Although the impacts of the policy are debatable (Shoko, 2002), it is important to note that the government was in fact trying to achieve environmental sustainability through addressing livelihood needs. When the country's economy collapsed the government used its legislative and enforcement power to shift towards a more restrictive mining environment, which resulted in small-scale miners being unable to carry out their activities openly. This reduced livelihood opportunities for the poor. This policy shift occurred at a time when the national economy was collapsing, and as shown by studies (UNDP, 2008), livelihoods were at their most vulnerable. In fact Maponga and Ngorima (2003) show that in the post-independent era the performance of the national economy has been the main determinant of the scale at which gold panning takes place, with surges in the activity directly linked to a poor performing economy. Criminalising gold panning therefore closed a livelihood opportunity which households had resorted to in the past. An unintended consequence of this policy shift was that it destroyed the possibility of local actors carrying out gold panning in a manner which did not threaten the physical resource base. The resultant damage to the environment, as in the Zhulube micro-catchment, supports the argument for achieving environmental sustainability through measures which ensure livelihood sustainability. From the above it appears that there is a strong relationship between policies, household activities and environmental sustainability.

Environmental insecurity, which is, among other things, characterised by soil erosion, deterioration of cultivated lands, degradation of rangelands, has often been blamed on the activities of the poor. One can argue that combining Blaickie's political ecology and the concept of environmental security provides a potent tool which facilitates an understanding of political-economic processes which affect the poor, and the 'rationality' of the responses of the poor. To the poor, and indeed to humanity in general, livelihood needs take precedence over protection of the physical environment. This is supported by the finding of this chapter that the poor are willing to protect the environment if that will somehow contribute to their livelihoods, even for a limited time.

On the basis of the evidence presented in the chapter one can also question whether it is possible for the Millennium Development Goals, particularly Goal Number 7 on environmental sustainability, to be achieved if livelihoods of the poor remain precarious.

Chapter 6

Sowing seeds of hope: the case of conservation agriculture in the smallholder farming sector

6.1 Introduction

Agriculture in southern Africa is mostly rainfed, and is dependent on rainfall that is largely unreliable in terms of total amount received and its distribution in time and space. As a result, agricultural production in general and crop cultivation in particular is insecure. Although irrigation is a possible solution to the challenges posed by rainfed crop production, its development is largely limited by, among other things, the huge development and operational costs that are involved[29]. Out of the approximately 40 million hectares that can be irrigated in the region, only 15% is actually irrigated (Sijali, 2001). Increasingly soil water management is being seen as potentially able to solve crop production challenges faced by smallholder farmers who face the twin challenges of unreliable rainfall and limited access to irrigation. However, an emerging question is, can field level water-resources management contribute towards improved livelihoods in the smallholder farming sector?

The dependence on rainfed agriculture has given rise to widespread food insecurity. The problem is that, because of historical reasons, rainfed agriculture in Zimbabwe is practised in low rainfall areas. Three quarters of all communal areas[30] have low to very low rainfall[31]. This means that agricultural production for the majority of Zimbabwean smallholder farmers is constrained by rainfall, most of which occurs as intensive, convective storms that are characterised by very high spatial and temporal variability. The result is that there are incidences of severe crop reduction caused by frequent dry spells, which can last up to 21 days (Rockström, 2000, Barron *et al.*, 2003, Mutiro *et al.*, 2006, Mupangwa, 2008). This is illustrated by the fact that in general there is crop failure in 1-2 out of 5 years necessitating external interventions such as food aid (Hove, 2006). Other factors such as the cost and (un)availability of inputs, poor transport network, and weak markets also add to the problems. In Zimbabwe these factors have become more prominent in the mid to late 2000s because of the economic meltdown related to the

---

[29] For example Lankford (2010) observes that irrigation costs are at 'eye-watering' $10 000 per ha.

[30] See Chapters 3, 4 and 5 for detailed descriptions of what communal areas in Zimbabwe are and how they came about.

[31] Zimbabwe is divided into five agro-ecological zones, and most smallholder farming takes place in Regions 4 and 5, which receive on average less than 600 mm per annum (Mtsi, 2008). Although irrigation is a possible solution to the problem of low rainfall, smallholder farmers generally have little access to irrigation (Manzungu and Van der Zaag, 1996). Of Zimbabwe's 200 000 hectares of irrigated land (Manzungu, 2003) smallholder farmers' share is estimated to be less than 10% (Zawe, 2006; Mtsi, 2008).

political crisis that gripped the country. The impacts of government policy and the economic meltdown on agricultural production have generated debates. For example, the food shortages which plagued the country in the 2000s have been attributed to government policies (Richardson, 2005). However, this has been disputed by other scholars who posit that by the 2000s the commercial agricultural sector had moved from maize production to horticulture and other commercial crops, thus the political crisis which significantly affected the white commercial farming sector, did not cause the food shortages (Andersson, 2007). It is also notable that from the mid-1980s the smallholder farming sector in Zimbabwe accounted for more than half of the country's maize yield, which is the staple in the country (Andersson, 2007).

Despite the well-known impediments to rainfed agricultural production, considerable interventions have been made in an effort to improve crop productivity even in those areas that are considered to be marginal for rainfed crop production. As already hinted above, this is because of the limited choices that are available to farmers. Consequently one of the important foci has been on how best to use the available water. Various initiatives directed at improving the capture and concentration of rainfall in the field, which increases infiltration of water into the soil, have been tried (Nyamudeza, 1999). In the first two decades after independence these techniques were generally being promoted under the rubric of rainwater harvesting. However, these have been combined with other agronomic practices and promoted under the banner of conservation agriculture (Twomlow and Hove, 2006, Mupangwa et al., 2008, Giller et al., 2009). The main focus has been agronomy (soil-water relationships and soil fertility) (Twomlow and Hove, 2006; Mupangwa et al., 2008; Giller et al., 2009) and the tillage practices that are needed to realise these benefits. While the proposed interventions can technically increase crop yields (Rockström et al., 1998) questions have been raised about their suitability and relevance for smallholder farmers who face many socio-economic related production constraints in addition to the biophysical ones (Munamati and Nyagumbo, 2010). Nevertheless conservation agriculture has been promoted as 'one-size-fit-all intervention' (Giller et al., 2009), and has continued to spread in the Zimbabwean rural landscape.

The main objective of this chapter is to analyse the reasons behind the promotion of field-level water resources management techniques, and how these are being accepted by smallholder farmers. The following questions are used to guide the inquiry: *how and why is conservation agriculture being promoted among smallholder farmers? Which specific conservation agriculture techniques are being promoted among smallholder farmers, and why have these been accepted or rejected by the farmers?* These questions are particularly important given that, although doubts have been raised with regard to shortcomings of conservation agriculture, its techniques continue to spread within the country. Furthermore, the chapter asks: *what characterises the relations between state and non-state actors in the research and promotion of conservation agriculture? What are the implications of these relations on the promotion of conservation agriculture among smallholder farmers?* Answers to these questions are critical when one takes into account the fact that in Zimbabwe, historically state funded extension has been supported by state funded agricultural research. However, because of the economic challenges which the country faced in the 2000s, this scenario has changed. This will be discussed further in the chapter.

Data for this chapter were collected between November 2007 and May 2009 through key informant interviews and focus group discussions with farmers in Ward 1. Staff from non-state organisations, organisations undertaking research and promoting conservation agriculture, and government extension workers, were interviewed. Participant observer method was also used. The researcher took part in a workshop that was organised by the International Crops Research Institute for the Semi-Arid Tropics (ICRISAT) at Matopos Research Station, Zimbabwe. ICRISAT is a Consortium Centre of the Consultative Group on International Research in Agriculture (CGIAR).

The chapter firstly presents a discussion on two concepts, that of narratives and that of intervention. Narratives, which are defined as interpretive stories (Molle, 2008) are commonly used in the social sciences to shed light on social reality. In this chapter narratives are used as a way of capturing the social dimension of conservation agriculture which is often underplayed and drowned in agronomic discourse. Intervention can involve the re-organisation or introduction of new practices and/or ideas into local situations on the belief that these will improve existing conditions or practices. These concepts will be applied to help in the understanding of why conservation agriculture is being promoted in smallholder farming, how it is being received by farmers, and the reasons that explain the manner in which smallholder farmers are receiving the practices of conservation farming. After that, as background to conservation agriculture, an overview of smallholder agriculture in Zimbabwe from a historical perspective is given. The section highlights the main challenges the sector has faced and continues to face, many of which can be attributed to government interventions, both in the colonial and independence era (Bolding, 2004). The chapter then turns to examine and analyse how conservation agriculture is being promoted among smallholder farmers in Ward 1. The section describes the techniques being promoted, how they are being promoted, and the perceptions of smallholder farmers towards the techniques. The same section also reviews the benefits of conservation agriculture, some of which are contested. Thereafter a closer look at the relation between research and extension, which is emerging out of the promotion of conservation farming, follows. The section presents a case based on a workshop which was organised by ICRISAT so as to understand the nature of emerging relations between researchers and extension workers. Within the same section the emerging relations between NGOs and extension workers are discussed. The chapter ends with a discussion and conclusions.

6.2 Narratives and intervention in smallholder farming

A major difference between the natural and social sciences is in the nature of data the two collect and rely on to reach conclusions. A characteristic of the natural sciences is reliance on data that are quantifiable, but unfortunately while such data serve their purpose, they fail to capture the human element in decision making processes, which can be subjective and be based on parameters that are not measurable. Perceptions, or views,

97

which are among the data the social sciences collect, capture what the natural sciences otherwise miss. This observation has led to narratives, which are stories that give an interpretation of some phenomenon (Molle, 2008) becoming increasingly important in understanding phenomenon. Although narratives have commonly been used in the social sciences, they have become incorporated into inter-disciplinary studies as a way of making social reality accessible through words and stories (Hyden, 2008). In development studies, for example, narratives have become important as it has become apparent that social actors have their own understanding of events and processes, and these understandings have implications on, for example, policy outcomes (Rimmon-Kenan, 2006). Whether collected through surveys, interviews or observations, narratives are often perceived as mere storylines. This makes them to be perceived as being subjective, and as not being fact (*ibid*). Thus a common mistake is to associate narratives with the local level actors only. However, a frequently ignored fact is that policymakers, agents of development and other such powerful actors design interventions based on their own narratives. Such narratives, which are based on interpretations of reality or ideas, may eulogize the importance of a particular intervention, for example. In fact, it has been observed that narratives of agents of development are often given from a position of power and are an ideological position. This can lead to situations whereby interventions are promoted not on the basis of facts. Thus in reality often interventions result in competing narratives, those emanating from agents of development, and those told by actors at the local level affected by the interventions. Among other things, narratives have become important as they give some form of power to local level actors as they make their views known. In the context of this thesis, narratives will be used to understand the perceptions of local actors on interventions in agricultural water management.

Intervention arises out of the belief by agents of developments, be they state or non-state actors, that local situations, life worlds or ways of organising social life are no longer valid or somehow ill-founded and inappropriate, and hence need to be re-structured if development is to take place (Long and Van der Ploeg, 1989). By intervening in local situations, agents of development try to shape or re-shape existing social practice and knowledge and introduce new elements that either replace or accord new meanings to already established ways of doing things. Thus intervention refers to:

> ... some kind of material or organizational input or package from outside (or the 'world beyond') which is designed to stimulate the emergence of certain 'internal' activities geared towards the achievement of higher levels of production, income generation, economic 'efficiency', or the better utilisation of existing resources and the 'human factor'. (Long and Van der Ploeg, 1989: 230)

Intervention is both of ideological and practical nature. Its ideological underpinning is the assumption that the injection of external inputs can provide a better solution to the problems than those that already exist (*ibid*). This assumption is problematic because the intervention may be considered to be unwelcome by the 'target group' or 'recipients' as they may not be beholden to perceptions of reality of the interveners. As shall later be shown in this chapter this has led to material incentives or inducements being offered to farmers so that the intervention (in this case, the introduction of conservation agriculture) does not fail. These inducements are legitimate points of enquiry because they may help explain the behavioural patterns that have been associated with conservation agriculture in Zimbabwe. Incentives or inducements are an offer of something of value meant to influence or alter a person's course of action (Grant, 2002). Grant further observes that material incentives or 'carrots' are commonly used to encourage cooperation (Smith and Stalans, 1991) and have been used in such areas as education, health care and in the promotion of Integrated Water Resources Management (IWRM) in Europe (Warner *et al.*, 2010). When incentives are used in a voluntary transaction they raise no ethical questions (Grant, 2002). In this chapter it will be shown how the incentives that have been 'dangled' in conservation agriculture have played a role in the adoption of conservation agricultural techniques. It should be noted that in both Zimbabwe's pre-independent and independent eras incentives, both positive and negative (in some cases extreme measures such as force), have frequently been deployed in smallholder agriculture to encourage or discourage the use of certain agricultural techniques (Bolding, 2004). Although promoters of conservation farming present it as an intervention that can increase yields and eradicate food insecurity, this chapter argues that conservation agriculture is in reality just one of the interventions in smallholder farming in Zimbabwe. However, what is notable about conservation agriculture as an intervention is that whereas in the past the state was the main actor facilitating change, in the case presented in this chapter non-state actors are the main force behind the intervention.

6.3 Smallholder farming in Zimbabwe: an overview

6.3.1 Biophysical and policy environment

From about the 10th century A.D. crop cultivation was an important activity in Zimbabwe (Manyanga, 2006). The result was that before colonialism, indigenous people were self-sufficient in food and produced surplus to cater for drought years and even fed the white settlers (Page and Page, 1991). Success depended on technologies or innovations such as shifting cultivation and pyro-culture (slash and burn), both of which were characterised by minimum disturbance of the soil (Manyanga, 2006), which in contemporary language could be minimum or zero-tillage. Mixed cropping was common. There was also an attempt to ensure that the crops grown were suited to local agro-ecological conditions.

These practices were, however, disrupted by settlers when they colonised what is now Zimbabwe at the end of the 19th century. Driven by colonial interests to seize black-owned land, settlers considered agricultural methods of the indigenous people to be wasteful and destructive. As Western crop production practices were introduced to replace indigenous ones, monoculture was promoted instead of mixed cropping while tillage using the plough replaced minimum or zero-tillage. Notable among the proponents of what was seen as modern farming methods that needed to be extended to black Africans was Emery Alvord, an American missionary. Alvord was the Agriculturist in charge of the instruction of Natives in the Department of Agriculture having been appointed to the post in 1926 (Bolding, 2004). His mission was to 'improve' agriculture in the areas occupied by the black population so that the land's carrying capacity would increase. This would make it unnecessary for the colonial government to allocate more land to an expanding African population[32]. A legacy of Alvord's work is the widespread adoption and use of the plough in Zimbabwe (Bolding, 2004).

Interventions in smallholder farming were not only limited to the introduction of the plough, but extended to even the re-organisation of indigenous population. The colonial government seized prime agricultural land from the indigenous people and allocated it to whites (Rukuni, 2006). The result was that at independence about 4,600 white commercial farmers occupied about 14.4 million hectares of prime agricultural land, which represented about 38% of the country by area (Whitlow, 1985). About 820,000 African households occupied 16.3 million hectares of mostly poor land, which is about 45% of the country (*ibid*). Only 12% of the land considered to be good for agricultural production was in the hands of black smallholder farmers. Of all land considered to be unsuitable for arable agriculture due to poor rainfall and soil conditions, 60% was under smallholder farming (Whitlow, 1985). This to a large extent explains the low productivity among smallholder farmers during the colonial as well as the post-colonial era.

The state of smallholder farming at independence, which as shown above was affected by colonial policies, prompted the independent government to adopt pro-smallholder farmer policies. One example of the pro-smallholder farmer policies adopted by the new government was that of the ending of racially discriminatory access to extension services and credit (Rukuni, 1994a). The result was that the number of black communal borrowing from the government-financed Agricultural Finance Corporation (AFC) increased from 18,000 in the 1979/80 season to 78,000 in the 1985/86 season (Rukuni, 1994b). Fertiliser purchased by farmers in this category rose by 45%, which in turn contributed towards increased maize output by the same farmers (*ibid*). The government also increased access to markets by raising the number of Grain Marketing Board (GMB) depots in the communal areas from three in 1980 to thirty-seven by 1991. Such moves facilitated an increase in agricultural productivity by smallholder farmers. The country received the World Freedom against Hunger Award in 1988 in recognition of the improvements in smallholder farming (Mashingaidze, 2006). Despite this accolade, and the title of the breadbasket of southern Africa, the majority of communal farmers continued to face widespread food insecurity especially in the semi-arid areas (Rukuni and Jayne, 1995).

6.3.2 Agricultural research and extension in Zimbabwe

Agricultural research and extension is credited with playing an important role in improving agriculture in Zimbabwe. Research in agriculture was institutionalised with the establishment of the Department of Research and Specialist Services (DRSS) in 1948 (Tawonezvi and Hikwa, 2006). During the colonial period the DRSS was concerned with the testing of imported livestock and crops for suitability to the local environment and also spearheaded the development of new crop varieties. Notable successes of the research effort included the breeding of high yielding wheat and maize varieties suitable for local conditions. However, these technological advances were largely geared towards meeting the needs of white commercial farmers. In the light of this neglect of smallholder

---

[32]Among the changes were cultivating, manuring and the practice of crop rotation (*ibid*).

farmers, the newly independent state re-oriented research towards solving the problems of the smallholder farming sector. Over the years research has tried to address the needs of the smallholder farmers with various levels of success. Research has focused on the breeding of high yielding varieties and also tillage practices.

Extension agents were tasked with the communication of research results to farmers. Agricultural extension for black farmers dates as far back as 1926 when African Agricultural Extension Workers were first trained (Pazvakavambwa, 1994). These were to be general agricultural advisers in the Tribal Trust Lands which had been set up by the colonial government to accommodate Africans. At the time the government forced Africans to adopt agricultural practices it considered appropriate. Thus interventions were imposed on the smallholder farmers and extension workers were foot soldiers of the colonial government in the process. Upon attaining independence in 1980, the new government re-organised extension services. In 1985 the Department of Agricultural, Technical and Extension Services (Agritex) was formed to improve extension services in the smallholder farming areas (Makwarimba and Vincent, 2004). This move ended segregated extension services which had served white and black farmers separately. The extension worker to farmer ratio was improved to 1: 800 (Pazvakavambwa, 1994). However, the gains which the government made towards promoting extension in the rural areas were affected by the economic and political turmoil of the 2000s. Inflation and general economic collapse weakened the government's capacity to support research and extension. This, coupled with natural disasters of drought worsened the state of smallholder farming in the country. Food insecurity, for example, became one of the major challenges in the sector. This crisis in the smallholder farming sector forms the backdrop against which conservation agriculture as an intervention must be understood.

6.3.3 Smallholder farming in Zimbabwe: A continuing crisis

Although the post-colonial government took measures towards improving smallholder farmer productivity, the country has experienced food shortages at both the national and household level. Successive droughts, among other things, have contributed to a decline in maize production. For example, maize yields in 1994/95 dropped by 42% from the 1993/94 yield (Chenje et al., 1998) as a result of drought. Drought was also responsible for the drop in yields from 2.1M tonnes in 2000 to 0.5M tonnes in 2002 (Jayne et al., 2006). Hove (2006) observes that since Zimbabwe's independence in 1980 drought relief programmes have been implemented in most rural parts of the country on a more or less annual basis.

The precarious food situation cannot just be attributed to natural factors such as drought alone. Herbst (1990) for example, blames poor government policies for the plight of many smallholder farmers. Ironically, the government seems to agree that its policies have been wanting. A Government of Zimbabwe (2009) document identifies challenges to agricultural productivity (which are said to have contributed to food insecurity), as including:
- Limited access to finance
- Lack of fertilizers
- Inadequate extension service

- Inequitable access to productive resources
- Poorly functioning markets
- Underdeveloped infrastructure
- Risks associated with adverse weather
- Lack of security of tenure (which has reduced investment in agriculture)

It must be noted that this analysis came after the collapse of the country's economy in the 2000s and in the aftermath of the fast track land reform programme which worsened the situation. In the same period the government introduced programmes aimed at reviving the agricultural sector, such as Operation *Maguta/inala* which was launched in September 2005, and the Farm Mechanisation Programme that was launched in June 2007. It is not the intention of this chapter to analyse the successes or failures of these interventions. What is important to highlight is the dominant (or should it be called domineering) role the government accorded itself in agricultural production after following the opposite policy during the Economic Structural Adjustment Programme (ESAP) in the 1990s. It is also important to note that from early 2009 the government began once again to scale back on its dominant role in agriculture. In many ways this created a gap which the non-state sector has tried to fill. This forms the backdrop against which conservation agriculture was and still is being promoted.

The next section looks at conservation agriculture as a recent intervention in smallholder farming in Zimbabwe, its American roots, and the reasons it is being promoted. This will be followed by an examination of how conservation agriculture is being promoted and adopted.

6.4 Conservation agriculture: state of the art

6.4.1 History and development of conservation agriculture

Conservation agriculture arose out of the search for sustainable agriculture. The United States and some South American countries were among the first to promote conservation agriculture (Haggblade and Tembo, 2003). In the USA policies encouraging conservation agriculture and the adoption of its techniques, such as minimum tillage, arose out of the need to conserve the natural resource base and minimise the use of oil-based fuels which powered farm equipment. Successive droughts in the 1930s, which threatened to turn the country's farmland into a dustbowl, called for an urgent need for changing approaches to crop cultivation. Among the techniques which were adopted to counter soil erosion were minimum tillage techniques. Disturbing the soil as little as possible was seen as contributing towards soil conservation in that it did not loosen the soil, which would lead to soil erosion. Furthermore, increases in the price of oil, especially in the 1970s, made the need to reduce energy consumption on farms imperative. Minimum tillage was seen as one way of reducing energy used for crop cultivation. Similar reasons also accounted for the adoption of conservation agriculture in South America, in particular Brazil, which by the end of the 1970s, was producing as much as a third of its crops using conservation agriculture (Lutz *et al.*, 1994).

102

In Zimbabwe the focus on conservation agriculture arose out of the need to ensure food security, as well as improve the livelihoods of smallholder farmers. Non-state actors, through the donor community, championed the intervention. In addition to distributing food aid the non-state sector diversified into supporting agricultural production with conservation agriculture being one such initiative. Conservation agriculture is being promoted mainly by NGOs funded under the Protracted Relief Programme (PRP) which is a programme of the United Kingdom's Department for International Development (DFID)[33]. PRP began in mid-2004 at a planned cost of £18m. Its stated aim was to increase agricultural productivity among vulnerable households and thereby increase household food security. Within the PRP vulnerable households are described as including the poorest in communities, female-headed households, those looking after orphans and infected/affected by HIV/AIDS, and those without access to adequate resources such as land and water, draught power and labour (Jones *et al.*, 2005). Such households are more likely to be practising rainfed crop production and to be more food insecure than households with access to irrigation. Implementation of PRP was through 10 international NGOs and inter-governmental organisations such as ICRISAT. Zimbabwean-based (local) NGOs were also included in the programme as implementing partners. The programme was being carried out in collaboration with United Nations agencies including the World Food Programme (WFP) and the Food and Agricultural Organisation (FAO). PRP was borne out of the realisation that in the months leading up to the harvest households run out of food and require food aid. Calculations by the promoters of conservation agriculture are said to have shown that if each household received seed and fertiliser inputs sufficient for 0.2 ha the household could produce an additional 300 kg of cereals. Potentially this could provide four adults with food for up to five months, long enough to cover the food deficit experienced in the period leading to the next harvest.

Donors consider conservation agriculture to be appropriate for such farmers because they perceive its tenets to be simple to understand, involving minor changes to existing practices, not requiring huge capital investments, and being consistent with sustainable development. Conservation agriculture is also touted as being able to raise the income of farmers (Kassama *et al.*, 2012).It is also seen as being cost-effective and better value for money than large-scale food distribution which NGOs commonly carry out. It is argued that although PRP costs 12% more to implement than providing food rations to cover three seasons, it has the potential to deliver twice the benefits in terms of increased food supply in the long run.

6.4.2 Techniques in conservation agriculture

As already noted conservation agriculture is characterised by minimum or no mechanical soil disturbance, permanent organic soil cover (consisting of a growing crop or a dead mulch of crop residues), and diversified crop rotations (Twomlow and Hove, 2006; Giller *et al.*, 2009;Dube *et al.*, 2012). In Zimbabwe one of the pioneers of conservation farming is Mr. Brian Oldrieve of the River of Life Church. Mr. Oldrieve promoted conservation

---

[33] It is notable that Zimbabwe's smallholder farming sector has also received support from several other donors, including the European Union.

farming as 'Farming the Christian way,' and wrote a handbook on its practices (Oldrieve, 1993). The handbook promotes conservation farming not only from an agronomical standpoint, but also from a religious perspective. It is argued that farming should be in harmony with nature and the spiritual realm. The handbook suggests that this can be achieved by, for example, using minimum tillage methods so that there is very little disturbance of the soil. The use of manure is also encouraged since this is said to be in tune with nature.

Giller *et al*., (2009) observe that there are many different practices which fall under the rubric of conservation agriculture. This makes it difficult to come up with one strict definition. They point out that, for example, zero-tillage, which conforms to the conservation agriculture tenet of minimum soil disturbance, on its own does not constitute conservation agriculture because it has to be accompanied by other practices such as mulching. The same is also true for planting basins. One can therefore argue that conservation agriculture has simply brought together different techniques and combined them, and this result in the need for substantial resources. It is questionable if smallholder farmers can mobilise enough land, financial and other resources needed to practice such full-fledged conservation agriculture.

By far the most common technique in the study area was planting basins. Basins are holes which are dug using a hoe. Generally their length/width/depth dimensions are 15cm x 15cm x 15cm. They are spaced 90 x 60 cm for maize. For sorghum and millet the spacing is 90 x 30 cm, while for cowpeas and groundnuts the dimensions are 45 x 30 cm. ICRISAT, which as will be discussed more fully in the next section, is one of the leading organisations in the promotion of conservation agriculture, recommends that planting basins be dug from early August through October, before the onset of the rains. Planting basins are said to enhance the capturing of rain, making it available to crops. Rainwater collects in the basins during the early season rainfall events (October/November). Planting is then done from November through to December after the basins have captured rains at least once. Farmers are encouraged to spread crop residues as a surface mulch to protect soil losses early in the season and conserve moisture later in the season. Planting basins require a lot of labour in the first year they are dug. However, once prepared theory suggests that the same planting positions can be used repeatedly[34], which reduces the labour requirements. Basins are also said to make it possible for farmers to be more precise in fertiliser application. It is recommended that instead of broadcasting fertiliser farmers should practice micro-dosing. Micro-dosing is when a farmer applies basal and top dressing fertiliser in the basin. ICRISAT recommends that a bottle cap be used to measure the amount of fertiliser to be applied. Top dressing fertiliser is applied between 3-6 weeks after crop emergence. Figures 6.1 and 6.2 show a field with planting basins, and size of basin.

---

[34] However, it is most likely that in sandy and other loose-structured soils planting basins need to be renewed more frequently than in clay soils.

Figure 6.1 Field with planting basins   **Figure 6.2 Size of planting basins**

Another technique which is also being promoted in the ward is that of dead level contours. This technology arose out of the modification of contours, which were designed to drain excess water away from the field (Van der Zaag, 2003) and thereby prevent soil erosion[35]. Dead level contours are designed not to channel water out of the field but to actually retain it in the field so that it can be made available to crops. Figure 6.3 shows a field with a dead level contour.

Figure 6.3 Field with dead level contour

In Manama area, which is a rural area in Gwanda District, it was observed that the 'dead level contours' on most smallholder farmers' fields did not achieve a zero gradient. Farmers claimed that they did not have the right equipment to guide them to construct proper dead level contours and that explained the slopes on the contours. Despite the

---

[35] Although the colonial administration forced Africans to use contour ridges as a soil conservation measure, these structures had an unintended consequence of connecting up a whole field and draining it into a single drainage line. This resulted in more powerful flows which had the power to erode (Wilson, 1995:288).

challenges which farmers in Manama faced in constructing dead level contours, some of them were invited to demonstrate to farmers in Ward 1, Insiza District, how to construct this type of contours.

6.4.3 Contested benefits of conservation agriculture

The attractiveness of conservation agriculture to donors is that, at least at the theoretical level, it is seen as addressing different challenges at multiple scales, namely at the field, household and watershed levels. At the field level the techniques are seen as addressing soil hydrology and fertility issues while at the household level conservation agriculture is seen as having potential to increase agricultural productivity among vulnerable farmers. At the watershed level conservation agriculture is claimed to contribute to sustainable utilization of natural resources. In general conservation agriculture is purported to increase soil organic matter content, increase soil fertility and reduce soil erosion (Giller et al., 2009). It is claimed that under conservation agriculture sedimentation of rivers and small reservoirs is reduced, which benefits the watershed.

As mentioned earlier, one of the limiting factors to crop cultivation under rainfed systems in semi-arid areas such as Mzingwane is soil moisture. Conservation agriculture is argued to contribute to increased crop productivity by increasing the availability of soil water to crops, which will in turn result in increased food security among smallholder farmers. It is also argued that conservation agriculture does not require much capital investment. This makes it suitable for vulnerable smallholder farmers, for example, those without draught power. Mulching is encouraged because it is said to improve water infiltration and reduce soil evaporation. This makes more water available to crops. Conservation agriculture thus in the short-term contributes towards household food security by increasing crop yields. In the long run it is said to contribute towards the conservation of natural resources.

However, there is no consensus among agricultural scientists as to how conservation agriculture impacts on crop yields (Mupangwa, 2008; Giller et al., 2009). Twomlow and Hove (2006) conclude that basin tillage benefits accrue to almost all the farmers using the technology. In 11 of Zimbabwe's districts all farmers who adopted the conservation agriculture technologies reportedly obtained significant yield gains from this technology (ibid). Their study showed positive yield gains in both below and above average seasons. However, Mupangwa (2008) paints a different picture. Using long-term assessment of the basin system through simulation modelling, he found that planting basins only gave marginal yield benefits over conventional tillage practices regardless of the nitrogen levels used. His work also showed that crop failure can be experienced in both conventional and basin systems in seasons with uneven distribution of rainfall. One of the conclusions Mupangwa makes is that food deficits are likely to continue as the performance of new technologies is dependent on the seasonal rainfall pattern[36]. Mupangwa's findings are closest to the conclusions of Giller et al. (2009) who suggest

---

[36] The question whether these practices work or not is also made difficult by that earlier work in Zimbabwe showed that different soils responded differently to the use of tied furrows as water conservation techniques (Nyamudeza, 1999). On soils with a 'reasonable' clay content tied furrows were said to be work well. However, a disadvantage was in a wet year tied furrows could result in waterlogging (ibid).

that short-term yield effects, which to a large extent determine the attractiveness of conservation agricultural practices to farmers, tend to be variable.

One reason why smallholder farmers may fail to reap the full benefits of conservation agriculture is because the technologies demand an 'integrated' approach to farming, which might be beyond the capacity of smallholder farmers to implement. Munamati and Nyagumbo (2010) add that there is little understanding among researchers of the socio-economic factors that can facilitate or hinder smallholder farmers from reaping the full benefits. Some of the 'must-dos' for farmers include weed control, application of fertilisers (organic and inorganic), mulching and crop rotation. Apart from these factors, Twomlow and Hove (2006) found that the starting quality of the soil and incidence of pests and diseases strongly influences the success or failure of the technologies.

6.5. Practising conservation agriculture

6.5.1 Farmers' views on conservation agriculture

This section presents a case on conservation farming in Ward 1, which is one of the wards in Insiza District in which it is being promoted among smallholder farmers. At the centre of the conservation agriculture intervention in the ward are non-state actors who are actively promoting it, World Vision being the main actor. All farmers who have received training from the NGO use planting basins. Incentives, mainly input packs of seed and fertilisers, are being used to encourage smallholder farmers to adopt conservation farming. This has worked to entice farmers into using basins. However, when the input schemes are discontinued, many stop practicing conservation agriculture. Some of the drop-outs reportedly reverted back to conventional ploughing. Interviews with both farmers and staff from ICRISAT and World Vision confirmed that it was common for farmers to plough over planting basins, for example, when they were left out of the inputs programme.

As far as the benefits of the planting basins are concerned, farmers stated that when the rains were below normal the basins were beneficial. However, when the rains were above average basins tended to collect too much water resulting in water logging. This was the case in the 2008-2009 season which had above average rainfall. Basins were also associated with the problem of mice eating seeds, resulting in low germination rates. Both male and female farmers complained that planting basins were labour intensive because of the digging and the weeding involved. This resulted in farmers in the ward being encouraged to form groups so that they could pool labour and share the workload. Some farmers in the ward said they had formed such groups (see below). Box 6.1 provides the views of four farmers on conservation agriculture:

Box 6.1: Perspectives of smallholder farmers on conservation agriculture

Mrs. Sibongile Ndlovu

"I started practicing conservation agriculture in 2004 after receiving training from World Vision. I do not have any cattle, so I 'plough with my fist' (with a handheld hoe). Basins are better for me. In 2006 we formed a working group of five households to help each other out with the labour, the digging of the basins and the weeding. However, the group collapsed in 2007 because some members were sending children to work on their behalf. In some years I get more from the plots with basins than from the plots without. In the 2005-06 season I harvested 5 x 50kgs bags of grain (maize). In 2006-07, and 2007-08 there were droughts. As a result I did not harvest anything. All the maize I harvest is for household consumption, I do not sell any maize. When I began using basins I received seed and fertiliser, but I have not received any since 2006.There is no farmer-to-farmer training taking place here. We learn from World Vision. Different people use different methods to fertilize their fields, some use fertiliser, some manure and others use soil from anthills."

Mr. Ncube

"I use planting basins because I see others do it. At one time everyone in the village wanted to plant in basins, but later when it came to selecting those to receive seed, only one household per village was given seed. This discouraged people. I don't know if my wife wants to use the basins, but as for me I no longer have the energy. I was disappointed because I was left out of the programme. What happened was World Vision selected people to receive seed and fertiliser, but the criteria used to select them were not clear and as a result some people in the village were disgruntled and left the programme. The amount of grain you harvest depends on the rainfall. The soil is also important. Basins are promoted because it is said if you use them crops will not wilt."

Mr. Matutu

"Some of my crops are grown in basins, some in the ways we have always been doing. I tried planting basins but the problem was that of mice. Some people are still using basins, but others have abandoned them. People were invited to get seed, but people from this village failed to get any and as a result some of those who had basins have dropped out. No one is sure about the criteria used to select those who received seed. If you do not have manure and use basins you do not get anything. Those with few cattle do not have manure. We do not see the AREX (Agriculture Research and Extension) officer, we just see World Vision. I think conventional methods are better because you can harvest more. You can only put 2 seeds in each basin."

Mr. M. Ndlovu

"I use both conventional methods and basins. The area I have ploughed in conventional methods is much bigger than the area under planting basins. The area I am cultivating is a new field and if I use basins I will have the problem of weeds. We learnt how to use basins at a workshop which was held at the village head's homestead. About 50 farmers were trained, of which 10 were men. The workshop was sponsored by World Vision. We received inputs from World Vision - 2 people share 50kg of fertiliser (top dressing); and 10kg of seed."

Among the smallholder farmers in Ward 1 it is common practice to use both the conventional and conservation tillage methods. Farmers stated that they do this as a form of spreading risk: if one plot fails at least the household would harvest from another plot. However, having a field under conservation agriculture also ensures that the household remains within the free inputs scheme. In almost all cases farmers interviewed stated that the field under conventional tillage was bigger in size than the field under conservation farming.

Apparently smallholder farmers are willing to be part of conservation agriculture networks in the hope that such networks can give them access to other projects run by NGOs and help them secure livelihoods. This view was confirmed by Mr. Mutero, who is a Scientific Officer with ICRISAT. Box 6.2 presents his views on the impact of incentives on smallholder cooperation with non-state actors.

Box 6.2: Consequences of using incentives in smallholder farming – the view of an ICRISAT officer

"Farmers are a bit tricky these days. Whenever they hear that there is a meeting organised by an NGO they are likely to come. That is why politicians are angry because when they call for a meeting people do not come, but if an NGO such as World Vision calls for a meeting people do come. Farmers know that if they attend an NGO meeting they will get something. This is why even if they know there is too much labour involved in the (conservation agriculture) technologies they still say they are able to carry out the work because then they get the seed packs. A very big question is, when these organisations go, what is going to happen to these technologies, to these things we are trying to promote? Is there any life after they go? If you go to Zhulube in August and September and tell people you can't get seed and inputs they (smallholder farmers) complain. What NGOs are now saying is that farmers who were in the programme from the first stage in 2004/2005 we are now weaning you off, look for your own inputs and continue with the technology. This season World Vision is working with new groups. Those who started the programme in 2004 and have been dropped, most of them don't care now. They don't use the (conservation agriculture) technology anymore. In one area in Masvingo the lead farmer actually said he was going to yoke his oxen and plough over the basins because he was angry he was no longer getting seed and fertilizer. Although his yields had improved and he acknowledged the success of the technology he was going back to conventional farmer practice because it is easier. Farmers put up the extra effort to deal with the technology because they get something from the NGOs, so if they withdraw than they can as well go back to their old practices."

6.5.2 Research and extension and conservation agriculture

This section analyses the interface between research and extension in conservation agriculture, specifically how non-state actors are working together with government extension workers. The analysis is based on the case of a workshop which was held by ICRISAT with government extension workers. The main purpose of the workshop was to train extension workers to conduct trials in conservation farming and to get feedback from trials which extension workers were conducting on behalf of ICRISAT. In the DFID-funded conservation agriculture network ICRISAT's role was to ensure that practices being promoted were scientifically sound. Towards that end the inter-governmental organisation conducted research on the performance of planting basins as a tillage practice and on soil fertility management, among other things.

As mentioned earlier, in both the pre- and post-independent Zimbabwe, government-led research and extension dominated the agricultural sector. However, this dominance was affected by the economic challenges of the 2000s. As government-led research and extension declined because of the economic challenges, non-state actors, some of which traditionally concentrated on emergence relief, gradually increased their role in smallholder farming, and even took over extension services.

To get insight into the dynamics among the different actors involved in conservation agriculture among smallholder farmers the researcher attended a workshop at ICRISAT's Matopos Research Station from 4-7 March 2008. The workshop brought together researchers from ICRISAT and 19 agricultural extension workers (of which 10 were male) from the Department of Agriculture Research and Extension (AREX). The extension workers were from Masvingo province and the two Matabeleland provinces. The workshop had the following objectives: to impart theoretical and practical knowledge on conservation agriculture, and to get feedback from the participants on the field trials on conservation agriculture which they were conducting in their areas. The field trial experiments were designed by ICRISAT and the role of the extension workers in the research was to: recruit farmers for the trials, help the farmers implement the experiments, and collect data from the farmers. Among the questions which the convenors of the workshop wanted to guide the proceedings were: *why do planting basins work? In what soils do they work? And where do they work?* It was also stated that the workshop aimed at building the capacity of extension workers in conservation agriculture so that they would '*have confidence in the technology.*' This confidence was said to be important if the extension workers were to be able to check on what NGOs were doing but also to carry out trials of their own.

However, during the workshop reports from the extension workers revealed the extent of the challenges which were making it difficult for the trials to yield relevant data. Extension workers complained that NGOs were causing confusion in the smallholder farming sector. Lack of coordination among the NGOs involved in conservation agriculture resulted in some smallholder farmers changing the NGOs they worked with almost each season. This made systematic data collection from the farmers difficult since different NGOs promoted different conservation agricultural packages. This was said to be made worse by AREX lacking the capacity to monitor the NGO activities and frequently not being aware of which technologies were being promoted and how.

For those trials that the extension workers were managing, the research focus was on impacts of different methods of managing soil fertility and land preparation techniques on crop yields. The trials sought to compare yields on fields treated with manure with those treated with fertilizer, and to compare yields on fields under conventional tillage with those on fields under planting basins. However, from the presentations it became clear that the extension workers had either not been instructed well on how to set up the experiments, or they had simply set up the experiments in a manner convenient under the circumstances they faced. Some of the problems which made the trials difficult to conduct in the manner designed by ICRISAT included shortage of land among smallholder farmers (which meant that not all trials could be conducted on land owned by a single farmer), and farmers changing the agreed tillage practice, cropping patterns and soil fertility management. Some farmers were said to be complaining about the whole idea of 'wasting' land by having fields devoted to experiments. The changes to the experimental design which were made by the farmers made it difficult for the extension workers to interpret and compare the data. Another challenge, which can be said to be originating with the extension workers themselves, was that of recording findings from the trials. None of the extension workers had exact yield records from the trials, thus, for example, findings were presented as: *"Fields prepared by ox-drawn ploughs did better than fields under basins."*

By the end of the workshop it was still not clear whether basin tillage outperformed conventional ploughing, or whether planting basins could be recommended for certain soil types or not. These observations are worrisome given that smallholder farmers taking up conservation agriculture were doing so based on the recommendations of non-state actors who at the time clearly did not have data to substantiate their recommendations. One can argue that smallholder farmers who adopted these techniques were actually risking their livelihoods by adopting techniques which could potentially fail, leading to loss of harvests.

The programmes that aimed at introducing conservation agriculture relied on a kind of division of labour between state and non-state actors that influenced the relationship between these actors. Earlier discussions with the agricultural extension worker in Ward 1 had revealed that she was basically coming in as and when World Vision wanted her help. She did not have any form of transport, which constrained her movements within the ward. This made it difficult for her to have her own programme on conservation farming. To investigate further the relation between state and non-state actors in conservation agriculture, Mr. Paradzayi, an extension worker based at Neshuro, Mwenezi District, which falls within the Mzingwane Catchment, was interviewed.

Mr. Paradzayi's view was that AREX was supposed to provide technical assistance to NGOs. However, what was happening in practice was that NGOs were simply doing as they pleased. At times NGOs simply went to the District Administrator's (DA) office to inform that they had a project they wanted to carry out with farmers. In such cases the DA would refer them to AREX as the relevant government department to deal with. He said NGOs would then come to AREX to inform them of the activities they were intending to carry out. In some cases NGOs would inform the department that a training workshop was going to be held with farmers. One NGO reportedly came to their offices and simply said they were going to train farmers in conservation agriculture and asked AREX "to organize the farmers". From then on that particular NGO never came to report on what they were doing. However, AREX learnt that about a hundred farmers had been recruited onto the project. NGOs were said to be only informing AREX of their activities so that they could report that they were running their projects together with AREX. This gave the activities of NGOs some legitimacy. Mr. Paradzayi also said that although AREX staff were not happy with the manner in which things were being done, there was nothing extension workers could do. In fact AREX staff appreciated the allowances which they received from NGOs whenever they were involved in their activities. The involvement of AREX in decision making on programmes and projects run by NGOs was limited because NGOs would bring programmes which had already been designed and packaged elsewhere. The type of activities to carry out, the number of farmers to recruit, the budgets involved and so forth were already decided by the time AREX got to know about the projects. Consequently it was difficult for AREX to influence these processes.

6.6 Discussion and conclusions

This chapter has discussed the promotion of field-based water resources management techniques in south-west Zimbabwe, which aimed at increasing agricultural productivity and improving food security at the household. The chapter relied on the narratives of different actors to understand two main aspects of conservation agriculture, *namely how it is being promoted and received among smallholder farmers, and how its promotion is affecting the dynamics of relations among the different actors*. This discussion focuses on these two aspects.

A major finding of the chapter was on the use of incentives to encourage smallholder farmers to adopt conservation agriculture. The chapter showed that non-state actors have to date been providing inputs which smallholder farmers need to practice conservation agriculture. Farmers, although they have doubts about the utility of conservation agricultural practices, such as basin tillage, accept the inputs that come tied to the adoption of its techniques. This can to a large extent be explained by the economic environment which prevailed in Zimbabwe in the 2000s. At the time the country was afflicted by a severe economic crisis whose impact also affected the cost and availability of agricultural inputs. It can therefore be argued that in a way the provision of inputs by NGOs filled a gap which smallholder farmers were facing, and therefore was a critical factor in the adoption of conservation agriculture. As the economic crisis was showing no signs of abating, for farmers it also made sense to be linked to the NGOs as a long-term survival strategy. This was so because the increasing role of NGOs in the food sector gave farmers hope that by adopting conservation agriculture they would stand a better chance of receiving aid (in its many different forms) from the same actors. Thus adopting conservation agriculture was a strategic move by smallholder farmers to meet an immediate need, that of inputs, and future aid related needs.

Conservation agriculture is a package of practices (Giller *et al.*, 2009), and this study found basin tillage and micro-dosing to be the main practices which smallholder farmers were adopting. As far as the actual practices which were adopted are concerned it is important to be mindful of two important factors in smallholder farming, which are labour and the agro-ecological environment. To understand the labour related factors one has to reflect on how the sector changed after colonisation, and is being changed by conservation agriculture. One of the innovation introduced to peasant farming in the then Rhodesia around the 1930s was the use of the plough (Ranger, 1985). Since then crop cultivation in the smallholder farming sector has been synonymous with ploughing. The adoption of the plough was linked to many factors, among them being the huge positive impact the implement made on labour needs. The plough enabled peasant farmers to cultivate more land with less labour (Krammer, 1997). Furthermore, the use of the plough reduced the burden of weeding. Studies have shown that the use of the plough reduces the problem of weeds, in some cases by as much as 80% (Davis *et al.*, 2006; Department of

Environment Food and Rural Affairs, 2011). However, it is interesting to note that basin tillage, a conservation agriculture practice, in a way reverses the benefits brought by ploughing. Basin tillage imposes a huge labour demand on the farmers as the basins are hand dug using hoes. Smallholder farmers who adopted conservation agriculture also complained that basin tillage was associated with the problem of weed control. These factors explain why smallholder farmers were found to be practising conservation agriculture only on smaller pieces of land when compared to land under conventional farmers practice. However, still there is need to understand why farmers did adopt basin tillage despite the fact that it increased labour needs and reduced the amount of land they could cultivate. One explanation, which forms the basis of the promotion of the practice in the sector, was that basins increased water availability to plants. Farmers were divided over this issue, with some stating that they had benefitted from the use of basins while others were sceptical. Controversy over the utility of basins was also found in literature (see for example, Giller *et al.*, 2009; Mupangwa, 2010). It is therefore possible that smallholder farmers adopted basins because of the hope they offered to farmers in a semi-arid environment. One can say farmers were willing to try anything that offered them hope of improved yields. However, as shown in Figure 6.2, well prepared basins are highly visible. This most likely worked to the advantage of smallholder farmers who saw taking up the basins as a means of proving their loyalty to non-state actors and therefore gain further access to donor projects.

In addition to basin tillage and micro-dosing, promoters of conservation agriculture encourage farmers to use crop residues as mulch. However, in the study area mulching was not observed. This can be linked to the agro-ecological conditions prevailing in the catchment. As mentioned earlier, the Mzingwane catchment is mostly dry (see Chapter 2), and that impacts on the quality of pastures. As a result farmers prefer to use crop residues to feed livestock to supplement the poor pastures rather than to use the residues as mulch. As was also mentioned in Chapter 2, livestock play a very important role in the livelihoods of smallholder farmers in the catchment. The fact that even those farmers who chose to practice basin tillage and micro-dosing did not mulch but used the crop residues for fodder shows the priorities farmers have. The agro-ecological conditions in the catchment made crop failure more likely than not, so using crop residues as fodder held better promises than using it as mulch. Arguably micro-dosing was adopted by the farmers since the fertiliser was provided by the non-state actors promoting conservation agriculture and therefore the farmers reasoned they had nothing to lose. In any case the concept of micro-dosing was that farmers apply fertiliser in small quantities (one bottle-cap per plant) which potentially could result in savings on fertiliser which could then be applied to other fields. As mentioned earlier, conservation agriculture is promoted as being about (green) water resources management at the field level. However, when one considers that farmers appear not too keen to adopt some of its practices (basin tillage and mulching), one can argue that the potential benefits arising from such water resources management practices have not been enough to lure farmers from their conventional practices. This position is supported by the fact that when the incentives, mainly farming inputs, stopped coming from the NGOs, smallholder farmers reverted to the conventional practice of preparing land using the plough. The fact that within a few seasons some smallholder farmers had been lured into conservation agriculture through incentives, and dropped its techniques as soon as they were removed from the free inputs scheme, raises serious questions about the sustainability of the approach used by NGOs to promote conservation agriculture. One can also argue that NGOs are merely trying to address the manifestations of structural problems affecting smallholder farming in the country. Inability to access inputs in particular, and agricultural productivity in general, are symptoms of deeper problems within smallholder farming. It is questionable if such problems can be solved through improved green water resources management at the local level.

As far as relations between state and non-state actors in the promotion of conservation agriculture are concerned, the chapter showed an almost patronage relationship between the two actors, with state actors appearing to being the dependent actors. Non-state actors were able to bring on board government extension workers to support or give legitimacy to conservation agriculture because of the financial resources that were available to them. State actors were brought in to contribute towards research on conservation agricultural practices, but they had little influence on the design of the research, nor could they halt nor encourage the promotion of conservation agriculture, given the findings of the research. Although some government extension workers had reservations about the usefulness of conservation agricultural techniques, they preferred to work together with the non-state actors promoting these techniques because of the financial rewards they received. One can argue that the state, weakened by a financial crisis and brain drain, could hardly regulate the promotion of conservation agriculture among smallholder farmers.

In closing it can be said that conservation agriculture as a field-level water resources management intervention is based on wide extrapolations and appears to have limited capacity to solve food-related livelihood challenges in the smallholder farming sector. As was shown in this chapter, conservation agriculture is being championed by non-state actors. Without the support central government it is doubtful if conservation agriculture can address the challenge of low crop productivity in smallholder farming (Nkala *et al.*, 2011).As an intervention it does not offer location-specific solutions which are necessary to solve the challenges of agricultural productivity in smallholder farming (Van der Zaag, 2010). In most of the study area rainfall is too low and erratic to allow the reliable production of grain crops. Moreover most soils are infertile and can only sustain maize production if expensive inorganic fertilisers are added. This is in addition to other factors that have to be brought to the equation, such as labour, draught power and markets. Given the variability of factors which characterise smallholder farming it is necessary that, for example, farmers be equipped to buffer against rainfall variability within a specific context (*ibid*). Against such a backdrop it can be asked whether it is possible, even under the best of circumstances, that smallholder famers can produce enough food to meet their needs on the back of conservation agriculture. Indeed it can further be asked whether conservation agriculture is the route to food security and agricultural productivity or alternative pathways have to be found, such as realising the importance of livestock production and off farm economic activities.

Chapter 7

Contestations and coalitions in urban water supply:
the state, the city and the politics of water in Bulawayo, Zimbabwe

7.1 Introduction

This chapter, just as the chapters that precede it, is about local water management practices and local responses to government interventions. However, the focus in this chapter shifts from a rural to an urban setting. The chapter describes the struggles of the city of Bulawayo, both its local authority and its residents, to gain and maintain control over the city's water supply. In the context of IWRM, the case of Bulawayo is of interest in that the central government attempted to re-centralise operational water management functions that it earlier decentralised. But this case goes far beyond the IWRM narrative. This is a case of the central government making desperate moves in the face of near economic and political collapse, and a coalition of local authorities and resident organisations that vehemently defend their control over water infrastructure. Their defence is not so much politically motivated but more of a fight for survival, for in this semi-arid environment the possible collapse of the water system would imply the end of the city.

This struggle for the control over Bulawayo's water services resonates with Swyngedouw's (2006) observation, albeit in a different context, that water is a 'hybrid' thing which captures and embodies processes that are simultaneously material, discursive and symbolic. Taken in this context water resources management entails, in one dimension, the management of a physical resource and the associated physical conveyance structures. Beyond the management of these tangible goods, water resources management is also about relations, socio-political and economic, which emerge through processes of capturing and conveying it, and also sharing it. Through such processes water acquires symbolic meaning. Where water resources management frameworks have been changed contestations have often been over the symbolic meaning of water. Water privatisation can be used to illustrate this. Fierce debates on water privatisation have been triggered by global institutions such as the World Bank, which have been advocating such an approach to the management of urban utilities. Swyngedouw (2006) suggests that contestations which have followed water privatisation have not so much been about whether water is managed by a private or public utility, but about local versus external control. In this chapter I show that many urban residents are keenly aware of investments made into water resources development and urban water networks by local authorities and feel connected to these investments and to the water so accessed. This hydraulic property investment defines the relation between the residents and water.

Closely related to the above is that changing the water supply services model redefines governance in society (Swyngedouw, 2002; Sneddon and Fox, 2008). When water supply services are privatised, for example, control of, and access to the service are impacted. The reverse, when water supply is brought under public management, also affects governance. Governance is affected by these shifts because they tilt the balance of power. In the case of privatisation power may end up concentrated in the service providing utility. In the same vein, encouraging decentralisation and participation through IWRM can be seen as processes of power (re)distribution. Using this argument, IWRM therefore is not merely a framework within which water is managed but is also about the empowerment of some stakeholders and the disempowerment of others. In this way IWRM can therefore be considered as a tool for statecraft. Following the formulation by Scott (1998) statecraft is understood as the process of state-making.

This chapter examines the struggle for the control of Bulawayo's water supply and sanitation (WSS) sector between the central government and the city, which in a way epitomises national politics in Zimbabwe. This struggle can be traced to the November 2006 Cabinet Directive which ordered all urban authorities to hand over their water and sewerage services to the national water authority.

Using the above mentioned Cabinet Directive as an entry point, the main question which this chapter tries to answer is, *how does context help in understanding contestations over the control of urban water services*? Specific questions of interest to the chapter are: *how do water coalitions emerge in water resources management*? And, *what role does the socio-political context and identities play in influencing the adoption or rejection of management frameworks?*

The chapter opens with a discussion on statecraft and governance. Secondly a brief background to the city of Bulawayo is given. Bulawayo's industry was at one time so well developed that it rivalled that of the country's capital city, Harare. The overview of Bulawayo will be followed by analysis of the case of the government's attempt to take-over the WSS sector in the same city. The basis on which the takeover was resisted and the response of local authorities and civic organisations is also described. Finally the chapter winds up with a discussion and conclusions. One of the discussion points is that water resources management must be looked at as encompassing more than the mere control of a physical resource.

Methodologically the chapter relied on documentary evidence such as minutes of the Bulawayo City Council, and correspondence between the council and state officials, and interviews with key informants from the Bulawayo City Council, residents' associations and non-state actors who were involved in campaigning against the proposed takeover. Fieldwork for this chapter, which was conducted between February 2007 and December 2008, was made difficult by that the issue within the state was only discussed at very high levels. Given the political tension in the country and sensitivity of central government at the time, this eliminated the possibility of interviewing senior government officials on the case. The research was conducted during a period which ranks as one of the most tumultuous times in the country's history.

7.2 Statecraft and governance

State interventions in resource management can be understood as a process of state making (Alexander, 2006). The state has often used such interventions to order, discipline and even to create identities and social relations (ibid). In Zimbabwe the Fast Track Land Reform Programme (FTLRP) of the 2000s is illustrative. Writing about land reform in the country, Alexander (2006:1) observes that, "This is not about the geography of displacement, expropriation, and re-apportionment...(but) the making and unmaking of authority over people and the land." Alexander concludes that the state is key in the locus for debate over the relationship between people, resources, power and authority.

The concept of statecraft by Scott (1998) fits in the ideas of the relationship between the state, citizens and authority. Scott looks at the activities of the state as attempts to create a society following some administrative and aesthetic order. Using examples of scientific forestry in Germany, ordered cities in medieval Europe and 19th Century France and villagisation in East Africa, he shows that the state has tended to use the legitimacy of administration or science and technology to justify its re-ordering of both nature and society. The state promotes its control (or withdrawal) on the basis that it improves administration, productivity or service delivery. Scott adds that the subtext of statecraft is in most cases political control. In the case of Tanzania, for example, villagisation was to re-organise communities to facilitate better political control. Scott observes that these 'high ambitions' of the state fail upon encountering reality, such as citizens who are opposed to it.

Coming back to the water sector, evidence of statecraft, in particular the re-ordering of nature, has for years been evident in government-funded dam construction projects. Allan (2003) refers to this as the hydraulic mission of industrial modernity. The re-ordering of nature was meant to increase agricultural productivity through increased water supply and to increase industrial productivity through the generation of electricity. However, as Allan shows, water resources management has over the years gone through sequences, each characterised by a particular stance of the state. The current management paradigm, which he sees as being dominated by water allocation and management, has seen the role of the state in the sector justified by the need for administrative efficiency and improved service provision. The role of the state here is taken to mean either direct state involvement, or the state facilitating other actors to come on board. The state can be directly involved in the administration of the resource, or indirectly influencing the management paradigm applied in the sector.

In the case of Zimbabwe, the water sector reforms sought, to an extent, to improve water resources management by shifting power from the government to stakeholders and semi-autonomous bodies. The ZINWA Act (1998) created the Zimbabwe National Water Authority (ZINWA) which was to be a parastatal operating along commercial lines (Makurira and Magumo, 2006; Mtisi, 2011). ZINWA was to have the responsibility of providing a coordinated framework for planning, development and management of water resources. In addition it was to raise funds by, among other things, selling, supplying and managing agreement water. The reforms also created catchment and subcatchment councils which were to be stakeholder participation platforms operating along hydrological boundaries. Catchment councils were given the task of preparing catchment outline plans for their respective areas, determining applications and granting water permits, regulating and supervising the use of water and supervising the performance of sub-catchment councils. Subcatchment councils were meant to regulate and supervise the exercise of water permits within their areas of jurisdiction, collect subcatchment fees and levies. They would also collect water levies, monitor flows and water use, especially in relation to allocations made by the catchment council. However, these ambitions of the government faced operational challenges. Funding constraints and lack of manpower capacity, for example, resulted in the newly created institutions being unable to perform their statutory roles (Makurira and Mugumo, 2006). To illustrate this one can point out that, although the national water authority was supposed to act as the secretariat to the catchment council, it ended up taking a leading role in executing the functions of catchment councils. In the case of urban water supply, a combination of the economic decline which affected the country at the turn of the millennium, and governance related factors culminated in the government handing over Harare water services to the national water authority in 2005 (Musemwa, 2011). Prior to this move ZINWA had been responsible for supplying Harare City Council, which had all along been managing the city's water services, with raw (bulk) water. This shift in responsibilities in effect added operational functions of purifying water and supplying it to residents of the city to the ones mandated to ZINWA by law.

The following sections give the background to the City of Bulawayo and analyse the attempted takeover of the city's water services. These will be looked at in relation to processes of statecraft in Zimbabwe.

The next section moves on to analyse the case of the struggle for the control of Bulawayo's WSS.

7.3 Background to the City of Bulawayo

With a population close to 650 000 (Zimbabwe National Statistics Agency, 2012), Bulawayo is one of Zimbabwe's major cities. It is also one of the oldest colonial settlements in the country having been established by white settlers in March 1894. On 1 June 1894 Bulawayo was declared to be a town by the colony's administration, and on 27 October 1897 it officially became a municipality with 3 wards (Matabeleland Chamber of Industry, 1994). Chikowero (2007) attributes the growth of the city to the availability of electricity. By 1924 electricity was being distributed around the city, which enabled industries to set up such that by 1945 the city was considered to be Rhodesia's industrial capital (ibid).

Water is a perennial problem in the city largely because of the low rainfall, with average rainfall being about 460 mm/a (Mkandla et al., 2005). The rains occur between October and March and are highly erratic and variable. The location of the city also contributes to the water problems which it faces. The city is located close to divide of the Limpopo and the Zambezi catchments, which means it is not close to the large rivers that can supply it water. All the rivers within easy reach are small and have small catchment areas (Bulawayo Engineering Services Department *et al.*, 2001).

7.3.1 Bulawayo's early days to 1980: a brief history of water supply management

Concerning water, the city's history has been characterised by the interest of the municipality to control urban water supply. According to Musemwa (2008) the monopoly of providing the town with water was granted to three pioneers. This concession was subsequently acquired by Willoughbys Consolidated Company which formed the Bulawayo Waterworks Company (BWC). Musemwa observes that at the time, within the colony, Bulawayo was the only settlement of its size which had its water supply in the hands of a private provider. The explanation which the colonial administration gave for that was that since Bulawayo did not have an effective sanitation board, water supply services had been handed to private providers as a way of expediting expansion of its water supply services. However, Musemwa states that BWC performance failed to meet the needs and expectations of residents. By 1910 the company was supplying water to only half the residents. Bolstered by the support of residents, Bulawayo's municipality began vigorous efforts to gain control over the city's water supply. On 1 July 1924 after court battles the municipality took over the services. Musemwa suggests that control of the city's water supply was more than just about water, but was in essence about political control over the city. The city wanted to rid itself of the constraints imposed on it by the British South Africa Company (BSAC)[37]. It was the BSAC that had granted BWC the water concession. To the city fathers and the residents, controlling water supply was a necessary condition for development. One can therefore conclude that even in the early days of the settlement water supply was a contentious issue which was not viewed just in the light of human needs, but also in the context of power, political control and development.

To deal with the perennial water shortages, the policy of the city between 1928 and 1976 was to construct a new dam every 12-15 years. This policy was meant to keep pace with population growth rates. This development strategy was discontinued after the 1976 Water Act took away the responsibility of providing bulk water from the local authorities and gave it to central government. The main dams which were constructed before this legislative change include Khami, Umzingwane, Inyankuni, Lower Ncema, and Upper Ncema dams. Table 7.1 shows the dams which supply the city with water and their capacities.

---

[37] The BSAC is the company which received the Charter from the Queen of England to colonise what is now Zimbabwe.

Table 7.1 Water supplies dams for Bulawayo

Dam	Capacity (Mm³)	River system	Year constructed	Owned by
Khami *	3.3	Gwayi	1928	BCC
Lower Ncema	45.4	Ncema	1943	BCC
Umzingwane	42	Mzingwane	1958	BCC
Inyankuni	80.7	Inyankuni	1965	BCC
Upper Ncema	18.2	Ncema	1974	BCC
Insiza	173.5	Insiza	1975 (stage 1) 1992 (stage 2)	79.5% BCC /20.5% government
Mtshabezi Dam	52	Tuli	1994	ZINWA

Source: Bulawayo City Council
*Khami Dam was decommissioned in 1988 because of high levels of pollution

7.3.2 Management of water resources in Bulawayo in the independence era

While the early days of Bulawayo stand out for the struggle to gain control of water services and the efforts to secure water supply to the city through infrastructure development, the post-independence era stands out for water demand management. In the 1990s this was clearly influenced by IWRM through international cooperation between the city and donor states. The city's water services fall under Bulawayo City Council's Department of Engineering Services. The department is also in charge of sewage reticulation in the city. Over the years the department has employed a three-pronged strategy to manage the city's water: managing demand, enlisting the support of residents in its programmes, and wastewater re-use. The operation policy of the reservoirs, which is in line with the Department of Water Development's policy on the operation of supplies dams, is that dams should hold at least 21 months' supply by 31 March of each year. If the quantity of water stored in the dams is not enough to last that long, then rationing is introduced. Between 1980 and 2000, water rationing as a water demand management strategy was resorted to in 1983, 1984, 1987, 1991, 1995, and 2000 (Bulawayo Engineering Services Department et al., 2001). In some particularly bad years, such as 1984 and 1991/92, water rationing was in force for about the entire year. During a water rationing period, normal practice is that household water consumption is limited to 600 litres/day, and the use of hosepipes for all purposes is banned. The rising block tariff structure was introduced in 1991 as a means of managing water demand. Consumption above a set threshold was charged at more than twice the normal tariff level. Another innovation of the city has been to re-use wastewater. There are about 180 connections to the reclaimed water system which supply water to parks and sports clubs The water is used to irrigate green areas (ibid; Gumbo, 2004). Waste water is also used for irrigation on some farms and to water road verges.

The approaches of the city to managing water demand have been quite successful. For example, Gumbo (2004) states that at the peak of the 1991-92 drought water consumption was reduced to about 45 000 m³/day, which worked out to less than 40 litres/cap/day. A direct effect of the stringent water rationing regime has been that water demand has remained suppressed despite the demographic changes occurring (NORPLAN AS Consultancy, 2001; Gumbo, 2004; Mkandla *et al.*, 2005). A study by NORPLAN AS Consultancy (2001) summarised the findings of water use patterns in the city in the following words:

> Bulawayo's water consumption patterns defy evaluation by conventional norms... Even during non-rationing periods, the Annual Average Daily Consumption does not return to an anticipated overall growth pattern. Rather it tends to follow the proportion of water available in the storage dams; almost as if the population of Bulawayo has reacted consciously or sub-consciously to the prevailing climatic conditions. (NORPLAN AS, 2001: 66)

Mkandla *et al.* suggest that the levels of water use in the city indicate that there is no need to aim for further lowering of per capita consumption.

A major component of the successes of the city in managing demand have been a result of the cooperation of the residents. Gumbo (2004) suggests that water demand management is 50% social engineering, which in the city has focused on encouraging water saving behaviour among the residents. The city has successfully managed to enlist the support of residents in its water resources management programmes through public campaigns and also through the use of civic organisations within the city. During the 1992 drought the city ran a massive media campaign which involved a weekly television programme, 'Bulawayo Must Live.' The programme informed residents about the levels of water in the city's supplies dams, dispensed tips on what residents could do to conserve water, and also informed residents of where they could report pipe bursts or get help on issues relating to water. The city council also keeps residents informed about the water situation in the city through pamphlets, posters and leaflets. In 2006, for example, the Department of Engineering Services produced a pamphlet which summarised the water situation in the city and urged residents to save water. These awareness campaigns have informed the residents about the water situation in the city and instilled in them a sense of responsibility as far as water is concerned.

After the 1991/92 drought, which was one of the worst in living memory, the city commissioned the Nyamandlovu Aquifer project, adding an additional and alternative (ground) water source to the supply system. A total of 68 boreholes were sunk into the aquifer and a 47 km pipeline laid to convey water to the city[38]. The drought also led to the formation of the Matabeleland Zambezi Water Project (MZWP) which advocated for the Zambezi River to be one of the city's new sources of water. Also as a result of the drought, in 1994 the government funded the construction of the Mtshabezi Dam. However, a pipeline to connect the dam to the city's water purification works is yet to be constructed, so its water remains unavailable to the city.

---

[38] At the time of the research less than half of the boreholes at the Nyamandhlovu aquifer were working.

7.4 The attempted take-over of Bulawayo water

Having given a brief history of water and the city attention now turns to the government's attempt to handover Bulawayo's WSS to ZINWA. This section will discuss the basis for the takeover, and the reasons for the refusal to handover the WSS sector to ZINWA.

On the 23rd of November 2006 a Cabinet Decision was made giving ZINWA the mandate to take over water and sewage delivery services in cities and towns with effect from 1st of December 2006. According to BCC officials, the first formal communication on the directive which the council received did not come from the Cabinet but came from ZINWA's Gwayi Catchment Council. The letter dated 02/01/2007 listed details which ZINWA required from the city council to facilitate the takeover, and these included: inventory of all customers, billing system, workshop, infrastructure layouts for both water and sewage by suburbs, water treatment works infrastructure including pumping plant, challenges being faced by the system, and the proposed 2007 recurrent and capital budgets on water and sewage.

The directive to ZINWA was justified on the grounds that water and sewer services in most urban areas of the country were deteriorating (Zimbabwe Parliamentary Debate, 2007). It was also reported that some urban authorities had actually gone to the extent of asking ZINWA to take over their water services. A Joint Task Force set up to investigate water shortages and sewer problems in Harare had also recommended that ZINWA take over these services. The position of ZINWA in providing water and sewer services was seen as being strong because the national water authority was already providing water to 'growth points'[39] and other small rural service centres in the country[40]. The government argued that the takeover would benefit the residents since funds for the development of the water sector would be channelled through one government department instead of being parcelled out to individual local government authorities. It was also envisaged that ZINWA would be able to use economies of scale and cross subsidies to cushion the poor in society from the effects of inflation on water charges. The government also argued that once the national water authority was providing water services in urban areas consumers would be assured of reliable, safe and sustainable water and sewage reticulation services.

Although the Cabinet used the argument that service delivery was declining to justify the takeover, one official suggested that the takeover was actually based on political reasons more than anything else[41]. He said that the loss of municipal elections by the ruling party's councillors in the 2002 had embarrassed the government. The ruling party lost elections in all major urban centres such as Harare, Mutare, Masvingo, and Kwekwe and

---

[39] Growth points are settlements earmarked for economic and physical development by the central and local government. The policy of growth points was started in 1981 as a way of redressing the colonially-induced development imbalances in the country (Wekwete, 1988).

[40] The *Zimbabwe Parliamentary Debates (2007)* report stated that ZINWA was providing water to 350 stations throughout the country.

[41] The official refused to be identified on the grounds that the issue of Bulawayo water was highly political and sensitive.

also in some smaller urban centres. The government's reaction to the loss was to dissolve the elected leadership of the city of Harare. The dissolved council was replaced by an appointed commission. However, the commission was failing to run the affairs of the city properly, shortcomings being particularly evident in the WSS sector. During the tenure of the commission, for example, water supply in Harare became very erratic with some suburbs going for several days without water. Such a situation was not only bad for the health of the residents, but also bad for the political image of the government and therefore could not be ignored. However, since the commission was the government's own creation, a second intervention into the affairs of the city had to be strategic and not seem as if the government was accepting failure. The government chose an approach which involved other urban centres where there were no appointed commissions. Taking over the WSS sector would give the impression that it was concerned about the welfare of residents. This approach would also give the government access to the coffers of the local authorities. Since Bulawayo City Council was dominated by opposition politicians, and the state perceived the city to be against it, taking over its WSS, and with it the revenue stream derived from the services, would cripple the city financially and ultimately contribute towards the control of the city by the government.

7.5 Resistance to the proposed takeover

Having looked at the proposed takeover of Bulawayo's WSS sector, attention now turns to opposition to the proposed takeover. It is tempting to start by analysing how the city council itself resisted the takeover. However, to show that even at the national level the takeover attracted a lot of interest, this section sets off by briefly looking at some of the debates that went on in Parliament concerning the matter. The section then moves on to show how the city council, together with civic organisations resisted the takeover.

7.5.1 Opposing debates in Parliament

One indicator that the proposed takeover of urban water services by ZINWA was not welcome was the manner in which it was debated in Parliament. When the Parliamentary Portfolio Committee on Local Government made its presentation to Parliament on the issue on 13 September 2007 a heated debate arose in the Senate. The presentation was made when ZINWA was already running the WSS sector in Harare, having taken over earlier in the year. The Senate questioned the capacity and performance of the national water authority in the city. It was argued that since ZINWA was already struggling to cope with the situation in Harare there was no need to continue with the takeover of the services in other urban areas. Another issue which raised concern was that the takeover did not get the benefit of input from the different stakeholders.

7.5.2 Resistance by the city council

Having looked at some of the points which were raised in opposition to the takeover of urban water at the national level, this section focuses on arguments which the council raised against the takeover, these centred on financial matters, legal facts and issues pertaining to council staff.

Financial matters as the basis for resistance to the takeover

The proposed takeover of Bulawayo's water threatened to cripple the city financially. The water account is the city's financial lifeline. In most years the water account generates a surplus which is used to subsidize loss-making services provided by the city. Apart from revenue from the sale of water, other sources of income for the city include Rates and Supplementary Charges, Licenses and Fees, and Rents. However, for these income streams the differences between expenditure and revenue is quite low because of the amounts the council spent in providing the service. Other accounts, such as the Health and Community Services Account and the Education Account, were said to consistently generate net deficits because residents simply stopped using the services when the fees became too high, yet the city could not shut them down. Thus, contributing over half of the council's income, the water account is very important to the city. Table 7.2 shows the contribution of the water account to the city's revenue.

Table 7.2 Contribution of the water account to Bulawayo's revenue

Year	% contribution
2004	54.4
2005	62.2
2006	67.6
2007	60

Source: BCC Minutes 22/02/2007

Figure 7.1 shows income and expenditure for the major accounts of Bulawayo City Council for 2006. There is a similar pattern of income from the water account, and the contribution of that account to the finances of Bulawayo City Council.

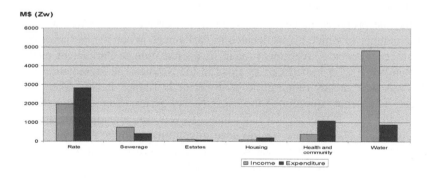

Figure 7.1 Income and expenditure of BCC for 2006

For the years shown, the water account did not only generate the highest income for the city council, it also had the highest surplus[42].

Related to this there were two other points of concern to the city council. One was the use of revenue collected in the city and another one was that of the details of the takeover. There was concern that once water supply was handed over to ZINWA revenue from the service would be taken out of the city and possibly subsidize ZINWA operations in other parts of the country. These fears were based on that ZINWA, because of its sole focus on water and having operations throughout the country, would probably not invest in other urban social services as the city council did. The BCC was also concerned that the takeover did not contain details of how the city council was going to be compensated for loss of income, and also for the investments it had made in infrastructure which was going to be taken over by ZINWA.

However, one should also hasten to add that at the time of the research the economic crisis in Zimbabwe was peaking, and inflation was estimated to be running in the millions. This economic environment was also impacting on the water services in Bulawayo. For example, the cost of chemicals needed to purify water was going up at such a rate that it was becoming almost unsustainable for the city to continue providing the service. In a 19/12/2007 memo to the Acting Town Clerk, the Acting Director of Engineering Services stated that the cost of importing Aluminium Sulphate was on average Zw$2,600,000/kg as compared to the local product last procured for Zw$268,575/kg in November. The change represented an approximate 870 % increase within the month. His opinion was that the costs were unsustainable. By the end of 2007 the water account became a drain on the income of Bulawayo, and on several occasions the city had to accept financial help from the state.

Exploiting legal loopholes to oppose the proposed take over

The city council also used arguments based on legal facts to oppose the proposed takeover. The main argument the city advanced was that the Cabinet Directive did not have a basis in law. In a letter from the Executive Mayor's Office dated 5 February 2007, BCC argued that the city derived the mandate to provide water and sanitation services to the residents from Acts of Parliament. Pieces of legislation which were cited include the Public Health Act (Ch 15:09) (Part VI) which puts a duty on local authorities to provide and maintain a sufficient supply of wholesome water for drinking and domestic purposes. That section of the law empowered the Council to run the water and sewerage reticulation system in the city. The section of the Act referred to states that:

> Every local authority, when required to do so by the Minister, shall provide and maintain, or cause to be provided and maintained as far as may be reasonably possible, a sufficient supply of wholesome water for drinking and domestic purposes,... and may construct, equip and maintain any works necessary for collecting, pumping or storing water.

---

[42] Accounts for years beyond 2006 were not made available to the researcher. For some of the years the accounts were still to be finalised. Currency changes such as the 'removal of zeroes' necessitated by the hyperinflation made it difficult to balance the accounts.

Another section of the same Act states that:

> All water works vested in any local authority shall be maintained by the local authority in a condition for the effective distribution of a supply of pure water for drinking and domestic purposes.

The council also referred to Part XII and XIII (Sections 168 to 190) of the Urban Councils Act which empowers urban authorities to carry out activities related to water and sewerage reticulation services. In a document entitled, *'City of Bulawayo Draft Position Paper on the ZINWA Takeover'* the council argued that existing legal and administrative frameworks governing water and wastewater management were clear and working well so there was no need to change them.

Staff matters and the proposed takeover

Closely tied to the legal arguments cited by the council were matters pertaining to the staff in the Department of Engineering Services. As was indicated earlier, the takeover of Bulawayo's water was planned in such a way that ZINWA would not only acquire the water and sanitation infrastructure, but would also transfer staff from the city's Engineering Services Department to the national water authority (Zimbabwe, 2007). Including staff in the takeover 'package deal' was a strategic calculation by the government which was meant to pre-empt the criticism that ZINWA did not have the capacity to handle water and sewerage services.

BCC argued that the proposal on the table was not workable since the concerned staff provided all engineering services provided by the city council, not just handling water and sewerage reticulation. Taking staff would therefore cripple the city's engineering department. BCC also argued that taking over staff could not be treated in the same way as taking over infrastructure. For instance, moving workers from BCC to ZINWA had implications on staff remuneration and motivation. On the issue of remuneration, ZINWA salaries were different (lower) from what the city council was paying its employees.[43]

7.5.3 Protest by civic society

Efforts to retain the control of Bulawayo's water services in the hands of the city council were led mainly by civic organisations. Among the civic organisations which were most active in the resistance against the takeover of the water were: Bulawayo Agenda, Radio Dialogue, Bulawayo United Residents Association, Bulawayo Progressive Residents Association, Media Institute of Southern Africa (Bulawayo Chapter), and the National Constitution Assembly (NCA) Bulawayo Chapter. What cannot escape attention is that some of these organisations were coming from opposite ends of the political divide to resist the government. For example, after the general elections of 2002 in which the MDC opposition party did better than the ruling ZANU P.F., the city's residents association, the

---

[43] Another dimension was that since ZINWA operated along hydrological boundaries, there was no guarantee that employees 'taken over' would remain stationed in the city. They could be posted to other parts of the catchment which would disrupt their family lives.

Bulawayo Residents Association (BURA) split into two. The Bulawayo Progressive Residents Association (BPRA) emerged from the split. BURA is perceived by residents to be aligned to ZANU P.F., while BPRA is perceived to be aligned to opposition political parties. Interestingly both associations were against the proposed takeover of the city's water services.

The fight for Bulawayo's water did not involve public demonstrations and protests. Civic organisations organised meetings at which officials from the city council were invited to give their views about the takeover, what the takeover would imply on service delivery, rates, and revenue for the city. One official with Radio Dialogue, which was one of the major actors involved in resisting the takeover, pointed out that they recorded interviews with city officials and other interested parties. At the time only the state-run Zimbabwe Broadcasting Corporation was licensed to broadcast in the country. To circumvent this restrictive situation, Radio Dialogue used a technique they referred to as 'Roadcasting.' By this interviews were conducted either in a studio, or at a public gathering, and then recorded on to compact discs (CDs). These CDs would then be distributed at street corners, bus termini and other public places. Basing their estimates on the number of CDs they distributed each month, the average household size, and the fact that CDs could be copied and passed along, the organisation estimated they were able to reach several tens of thousands of listeners each month. To increase interest in the CDs, interviews were interspaced with popular music.

Box 7.1 gives some of the views of the key actors who were involved in the protests, and these typify the common sentiments among the residents of the city.

Box 7.1 Voices of resistance

The Chairman of BURA:
> Whether you belong to this party or that one you still need water, we all need water. Politicians, churches and everyone is united on this one. This is not about politics that this one is this or that, but whoever has been in council the management of water has always been the best. *(10/06/2008, Bulawayo)*

The Advocacy Officer for Bulawayo Agenda:
> I do not remember anyone supporting the takeover. The people were saying the government killed people during the Gukurahundi and now it wanted to kill them through water. (25/02/2008, Bulawayo)

Minutes of the Bulawayo Agenda Meeting held on 22/02/2007:
> Past failures of parastatals are making the people sceptical. As is typical of all government controlled parastatals, ZINWA has recorded more failures than successes in the areas where it operates. The government has done irreparable damage to the city of Bulawayo, for example, the collapse of NRZ (railways); ZESA (electricity); CSC (meat) and industry in general." Arnold Payne: "What is national interest? It is a known fact that our 'democratically elected government' does not have the interest of people at heart.[44]

Programme Officer of the BPRA:
> What is it that Bulawayo City Council has done wrong, besides the fact that the councillors are all opposition? Actually I would understand it if Bulawayo was said to be taking over the running of everything in Zimbabwe. Bulawayo should be taking over the running of water and sewerage in Zimbabwe. (09/06/2008, Bulawayo)

Opposition to the takeover was also based on the perception that the state did not have the resources to run the WSS sector in Bulawayo. This perception was supported by economic indicators, particularly inflation and the shortage of foreign currency which was prevailing at the time. The perceived inability of ZINWA to improve service delivery in Harare also damaged its credibility to be capable of running un urban water supply system. ZINWA was also seen as having failed to carry out its core business of supplying bulk (untreated) water. At the time only 5 of the boreholes at the Nyamandlovu Aquifer were working[45], and the views of the people were that if ZINWA was serious about making a contribution to solving the water problems in the city it could start by repairing the boreholes under its charge.

Past experiences also helped to shape the perceptions that the government was not competent enough to take-over the WSS in the city. In 1989, for example, the state transferred the generation and distribution of electricity from local authorities to the Zimbabwe Electricity Supply Authority (ZESA). Bulawayo's thermal power station consequently was handed over to the authority, and residents complained that after that electricity problems of the city got worse.

---

[44] Punctuation within the quotation are in the original.

[45] Personal communication, Bulawayo City Council Principal Engineer, (10/06/2008, Bulawayo).

The proposed takeover was not effected and eventually in January 2009 the state ordered that ZINWA return the services it had taken over throughout the country. No explanation was given on the policy reversal.

7.6 Discussion and conclusion

The chapter presented a case on the struggle for the control of urban water services between central government and a local municipality. Case studies in the water sector have characteristically portrayed the tussle for the control of urban water services as pitching local and national authorities in one corner, and international financial institutions and private companies in the other. This analysis has its foundations in the privatisation of urban water services, which in most cases has been at the instigation of the World Bank and regional development banks and related institutions. However, in this chapter analysed a struggle for the control of urban water services between national and local authorities.

The chapter described the reasons behind the proposed takeover of Bulawayo's water, and the reasons for the resistance against it. The government argued that the city, just like others in the country, was failing to supply residents with domestic water. The government further argued that it would be cheaper and more efficient if the national water authority was to run the water supply system of the city since it was already providing similar services to other urban areas of the country. The reasons advanced by the government to justify the proposed takeover were based on issues of efficiency. Scott (1998) argues that such reasons are often used by governments as the smokescreen to hide statecraft projects. This proposed takeover was rejected by the Bulawayo City Council on the basis that the government had not proven inefficiency or mismanagement of the water supply system by the city council. The city council also argued that the proposed takeover was illegal since the city council was mandated by various pieces of legislation to supply water to the residents. Civic organisations sided with the city council in rejecting the proposed takeover.

However, to understand the struggle for the control of the city's water services one has to situate the struggle in the socio-economic and political context in which it took place. The proposed take-over was made at a time when the Zimbabwean government was under pressure politically and economically, from within and outside the country. Inflation, for example, was at a record high and this was causing massive suffering among the general population. Opposition political parties, which controlled some of the most important urban areas of the country, were calling for the government to step down. It can therefore be argued that the government's involvement in Bulawayo's water services was not just about the control of water, but to an extent was about the government trying to exert its influence in an urban area and regain control. The validity of reasons of efficiency which were used to justify the takeover is questionable given that the government was economically at its weakest. It would have made economic sense at the time for the government to relinquish services it could relinquish rather than to adopt new ones, such as urban water service provision. Analysing the same point from a reverse angle also suggests that the reason why the local authorities were not keen to let of the city's water services were not of an economic nature in the strict sense. Bulawayo city council, just like most other local authorities in the country, was also facing a financial crisis. As was shown in the chapter, the city was having difficulties procuring water treatment chemicals. On the basis of financial reasons alone, it would appear letting go the water services would have been a sound decision. However, the fact that even under severe financial stress the local authorities chose to fight the government to keep the water services points towards the fact that at stake was more than the control of water as a physical resource. This suggests that the struggle for the city's water services was in actual effect over the control of the symbolic meaning of water.

As Swyngedouw (1999) argues, water acquires symbolic meaning through processes of capturing it, conveying and allocating it. The chapter showed that right from the early days of the establishment of the settlement the city worked hard to gain control of its water supply and to secure its access to the physical resource. The city has a long history of dam construction, a policy which was only ended through changes to water law and policy. There is no doubt that among the major urban settlements in the country none has invested in water supplies dams as much as Bulawayo has. The century long struggle to secure water for the city combined with, and strengthened the shared identity that the people already had. Water thus attained also a symbolic meaning This is exemplified by the manner in which the campaigns to save water during the great drought of 1991/1992 were framed: "*amanzi yimpilo*" (water is life), and 'Bulawayo must live.' Arguably, controlling water, investment in hydraulic infrastructure and even 'wise use' of water has become part of the identity of the residents of Bulawayo, and a deep acknowledgement that without a secure and efficient water supply the city's future was in peril.

Bulawayo's water coalitions were a result of the physical environment of the region. Water scarcity necessitated investment in hydraulic infrastructure, which the city could not do without the financial support from residents. Investments in water infrastructure alone were not enough, they needed to be augmented by water demand management, so further support of the residents through water saving practices was also needed. The city therefore actively involved citizens to support its various water projects, for example, by running media campaigns. Such processes gained the support of residents, and in this way water coalitions were formed. Water coalitions also emerged from the civic groups which were springing up in the city and in the country in general to oppose the government. Among the civic organisations which opposed the proposed takeover were those that claimed to be championing democracy and good governance in the country. The city council successfully appealed to such organisations for support, further enlarging and strengthening water coalitions. An interesting feature of the coalitions was that civic groups which ordinarily were sympathetic to the government and those opposed to the government were united on this cause.

Although central government appeared resolute in its effort to take over the water services of the city, eventually the economic crisis in the country worsened to such an extent that the central government had to focus all its energies on the most urgent issues and that was most probably the reason why by January 2009 the takeover bid was cancelled. Even if that was the case, the role of the strong bottom-up movement that opposed the takeover cannot be ignored. Coalitions of civic groups defended the control over life-giving water infrastructure, irrespective of existing policies and local political divisions. It can therefore be said that the case presented in this chapter goes beyond the conventional IWRM narrative, and may in fact be a story of people who defend their identity and their future.

Chapter 8

Processes in river basin planning:
the case of the Limpopo river basin in Zimbabwe

8.1 Introduction

River basin planning, which in Zimbabwe is commonly referred to as catchment planning, has closely followed the socio-political history of the country in general. It predated the IWRM-informed water sector reforms that began in the mid-1990s and culminated in the promulgation in 1998 of both the Water Act (Zimbabwe, 1998a) and Zimbabwe National Water Authority (ZINWA) Act (Zimbabwe, 1998b), which adopted IWRM as the water management policy. The Water Act of 1976 (which remained in place 18 years into independence), made catchment planning a statutory requirement as did its successor, the Water Act of 1998. To understand whether the changes made to water law have, or can impact positively on livelihoods this chapter examines the process and content (outcomes) of catchment planning by seeking answers to the following questions:

> ➢ *what processes were involved in the making of the Mzingwane catchment plan?*
> ➢ *which aspects of water resources management did the catchment plan address, and how does this fit in with IWRM?*
> ➢ *how do the outcomes of catchment planning match the livelihood realities in the Mzingwane Catchment?*

Answers to these questions can help clarify whether river basin planning is conceptualised as development planning (which involves the prioritisation of the developmental goals of a society), or as a resource plan that promotes efficient water use. How river basin planning is conceptualised and executed potentially impacts on livelihood.

In the global water resource management discourse there has been a resurgence of river basin planning (Molle, 2009; Warner *et al.*, 2008). The World Summit on Sustainable Development (WSSD) of 2002, for example, called on countries to develop IWRM and Water Efficiency Plans[46]. It is this implied connection between the IWRM framework and sustainable development, in terms of whether river basin planning can and does contribute towards socio-economic development, especially among the rural poor, which interests this chapter.

---

[46] Other fora where IWRM has been discussed include the annual World Water Week which is hosted by the Stockholm International Water Institute, and the triennial World Water Forum, convened by the World Water Council, which brings together donors, academia and civil society.

The chapter focuses on how the river basin plan of the Mzingwane catchment, which is a sub-hydrological zone of the Limpopo river basin that lies in Zimbabwe, was made and whether it addresses the livelihood needs of the water users in the catchment. The Mzingwane catchment council is one of the seven catchment councils formed in Zimbabwe by the water reforms of 1998. It is one of the driest catchments in the country and is mostly dominated by smallholder farming. As has been indicated in Chapters 2, 4, 5 and 6 of this thesis, poverty is a major livelihood challenge in the Mzingwane catchment. A critical human need in the same catchment is that of improved access to water for domestic and productive uses as was highlighted in Chapter 4. Low agricultural productivity (discussed in Chapter 6 of this thesis) is also a challenge in the catchment. The empirical material is based on the draft catchment plan for the Mzingwane catchment produced in 2006, and gazetted and approved in last quarter of 2010[47].

The structure of the chapter is as follows: firstly the relationship between development planning and resource planning is explored. A brief history of river basin planning is given, focussing on how river basin planning has evolved over time. This is followed by a description of the research design and methods used to gather data for the chapter. The chapter then turns to Zimbabwe's experience in river basin planning. The Mazowe catchment, which is one of the country's wettest catchments and was a test catchment for water sector reforms, is used to trace how river basin planning evolved in the country. After that attention is directed to water resources planning in Zimbabwe's Mzingwane catchment. The process through which the draft plan for the Mzingwane catchment was made is traced. Among other things the section will analyse to what extent the planning process in the Mzingwane catchment was informed by experiences from the catchment planning process carried out during the pilot phase of the water sector reforms in the Mazowe. Thereafter the chapter compares and contrasts river basin planning as done by the Mzingwane catchment council with a planning process which was carried out in Ward 1, Insiza district, which falls within the Limpopo river basin in Zimbabwe. This will be done to see whether the two plans match, and if the plan made by the catchment council addresses livelihood needs of water users at the ward level. Finally discussion and conclusions of the chapter are presented.

8.2 Conceptual framework

8.2.1 Development and resource planning

The conceptual framework presented here focuses on development planning and resource planning, specifically on whether or how the two are related. Development planning, which is mostly carried out by central governments, has a broad outlook and focuses on a whole range of socio-economic issues, from education to social welfare, from health to infrastructure development, among other things. Development plans can be understood as political strategies of governments of the day, which reflect the ideological persuasion of governments in terms of how they interpret reality and set priorities (Filho and

---

[47] This chapter restricts itself to analysis of the draft plan not the final plan of the Mzingwane catchment which was eventually gazetted and approved. The plan which was eventually approved was made after data collection for this chapter had ended and so the process of finalising the plan was not studied.

Gonçalves, 2010). Through development plans governments make choices on what projects to embark on to address specific challenges or to create a particular state (Scott, 1998). Resource plans are usually prepared by sectoral ministries but approved by the finance ministry. Sectoral ministries produce plans regarding how different resources under their purview (such as water and minerals) can ideally be linked to and translated into development plans. Resource plans usually focus on a single resource and can have as their starting point, an analysis of the status of the resource (such as the physical state of the resource, for example), and depending on the nature of the plan, scenarios can be presented which show the possible ways of exploiting the resource. The likely developmental outcomes may or may not be presented.

8.2.2 Institutional models in river basin planning

River basin planning originated during the industrial modernity era (Molle, 2009). From the 1880s to the late 1970s river basin planning focused on water resources management, which, was characterised by the development of (huge) water infrastructure. This is commonly referred to in literature as the era of the hydraulic mission (Allan, 2003). The adoption of the river basin scale was encouraged by the perception that the river basin was ideal for regional development planning (Molle, 2009).

The adoption of the river basin scale for management and planning purposes is associated with the emergence of two main institutional models, namely the centralised/unicentric and the polycentric/decentralised/coordinative institutional model (Svendsen *et al.*, 2005; Lankford and Hepworth, 2010). The centralised model is characterised by the presence of a single unified organisation which makes decisions concerning the management of a river basin (Svendsen *et al.*, 2005). This concentration of power within one organisation enables the river basin authority to plan and implement projects. The Tennessee Valley Authority (TVA) of the United States of America is an example of such a model (*ibid*). Arguably, the TVA was able to support regional development through water resources management because its mandate at formation was to spearhead regional development. The coordinative institutional model is when different organisations, or layers of government, are coordinated to cover an entire river basin or sub-basin. Institutionally and organisationally this model is thus decentralised (Svendsen *et al.*, 2005; Lankford and Hepworth, 2010). The model depends on voluntary participation of different organisations/stakeholders, which can be an advantage since participation results in a strong political base (Svendsen *et al.*, 2005). However, participation based on goodwill can also stifle progress since levels of commitment to decision making processes maybe different among the stakeholders. Within the IWRM framework institutional decentralisation is encouraged, which makes it more aligned to the coordinative model than the centralised model. However, given that the coordinative model depends on the willingness of other sectors to cooperate, there is the possibility that delivering socio-economic development can be constrained by factors outside the scope of the river basin organisations.

In the post-World War 2 era, the TVA model (which is the centralised institutional model) became one of USA's exports to several parts of the world including Africa. It was thought that (integrated) river basin planning would make it possible to overcome impediments to development. In Africa TVA clones emerged with the formation of River Basin Development Authorities (RBDAs) in Nigeria (Adams, 1992). RBDAs planned for flood control, navigation, pollution control, fisheries, seed multiplication, food processing and livestock breeding, among other things. This comprehensive approach also gave rise to problems - grandiose plans were made but were not supported by financial muscle as was the case with the TVA (Adams, 1992). This limited the effectiveness of RBDAs. That the TVA was considered successful enough for it to be replicated in other parts of the world yet similar institutions in Africa failed suggests that there are limits to which institutional models can be transposed from one context to another (Shah et al., 2005).

IWRM-reforms have resulted in calls for a centralised institutional model of river basin planning to be abandoned in favour of what could be called coordinative models. Furthermore, within the framework the creation of resource-oriented rather than development-oriented organisations is advocated for. This makes the relationship between resource management and socio-economic development unclear. In Zimbabwe river basin and sub-basin organisations, known locally as catchment and subcatchment councils, were formed in line with the IWRM framework. These were supposed to bring together different stakeholders to the water resources management forum. In South Africa IWRM-based reforms resulted in the country being divided into Water Management Areas (WMA) which follow hydrological boundaries. Each WMA is under a Catchment Management Agency, below which are Water User Boards (WUBs). These organisations act as stakeholder platforms.

Scale has emerged as an important factor that influences the various water management institutions. For example, national level water management institutions operate within a regional development framework as represented by SADC[48] and basin framework represented by the Limpopo Watercourse Commission (LIMCOM). LIMCOM was established in 2003 to coordinate river basin management at the transboundary level (Fatch et al., 2010), and brings together the four riparian countries, which are Botswana, South Africa, Mozambique and Zimbabwe. All these countries are supposed to participate in the planning, utilisation, (sustainable) development, protection and conservation of the Limpopo river basin. The purpose of the commission is to coordinate water resource planning in the basin so as to avoid conflicts among riparian countries. Interestingly, there are currently no links between river basin organisations within the individual countries and LIMCOM (ibid). Each country is represented on the commission by a representative from the Ministry responsible for water as interactions among the riparian countries take place at governmental level. This breaks the links between lower tier organisations based in the individual riparian states, which are stakeholder-driven, and the basin commission. It is also notable that the actual development of water resources, such as water infrastructure, appears to be still organised through bilateral agreements between and among riparian countries, such as that between South Africa and

---

[48] The regional body, the Southern African Development Community (SADC), has crafted a number of instruments such as the SADC Protocol on Shared Watercourses which recognises the need to develop water resources and infrastructure for purposes of socio-economic development (SADC, 2005). The SADC Regional Strategic Action Plan makes reference to the fact that the development of water infrastructure in the region needs to be prioritised for purposes of delivering benefits to people (SADC RSAP, 2005).

Botswana (Turton, 2005). Bilateral agreements have been used as platforms for joint planning and development of water infrastructure in the basin. Figure 8.1 shows the formal institutional arrangements that are supposed to govern water management in the Limpopo river basin. What is not shown in the figure are non-water institutions which have a bearing on how water resource plans can be translated into developmental outcomes.

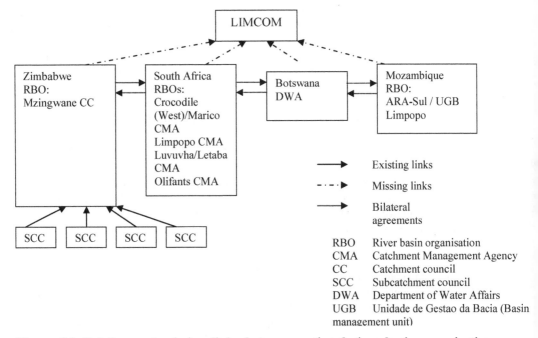

Figure 8.1 Existing and missing links between national river basin organisations within a transboundary river basin

8.3 Methodology

Data collection for the chapter was undertaken between April 2006 and May 2009. The study undertook to make a comparison between: two eras in water resources management in Zimbabwe (the pre-IWRM and the IWRM era)[49], catchment planning during the pilot phase in the Mazowe and catchment planning in the Mzingwane after the water sector reforms, and the interactions between various stakeholders at different scales over catchment planning. This was achieved by undertaking a literature review of the history of water resources management in Zimbabwe and in the Mazowe catchment in particular. The Mazowe has a long history of institutionalised water resources management in the

---

[49] Dividing the eras into colonial and the post-colonial era was avoided for the simple reason that the Water Act (1976) was only repealed in 1998 and therefore straddles both the colonial and post-colonial era. Thus it was felt that dividing the eras into pre-IWRM era and the IWRM would be clearer as the pre-IWRM has a more distinct end point.

country, and water sector reforms were piloted in the catchment. Further evidence was also collected through interviews with a former water supplies manager and the catchment hydrologist in the Ministry of Water and Rural Resources.

Data collection on river basin planning in the Mzingwane catchment included, among other methods, document review. Documents which were reviewed include minutes of meetings of the Mzingwane Catchment Council and the draft plan of the same catchment. Interviews were held with the Zimbabwe National Water Authority (ZINWA) staff operating in the Mzingwane catchment. Among the personnel interviewed was the catchment coordinator, who was directly involved in the process of making the Mzingwane catchment outline plan. The planning manager of ZINWA (Head Office) and the chief planning engineer in the Ministry of Water and Rural Resources were also interviewed. A chief, representing traditional leaders, was also interviewed to get an understanding of the participation of rural communities in catchment planning.

Interviews were also conducted with councillors from each subcatchment council within the Mzingwane catchment in order to determine the level of stakeholder participation and contribution to river basin planning. Since the Water Act (1998) stipulates that river basin plans must take into account regional development plans and any other plans, the researcher interviewed staff from Rural District Councils (RDCs). The RDCs which were selected for these interviews were Mberengwa RDC, Insiza RDC, Mwenezi RDC and Mangwe RDC.

The Mzingwane catchment is mainly rural and is presided over by several chieftainships. There are however, also a number of urban areas in the catchment. These include Plumtree, Beitbridge, Gwanda and Bulawayo (which is Zimbabwe's second largest city)[50] and Gwanda. Representatives of these centres were interviewed to determine the level of participation in catchment planning.

Water users at the local level[51] were included in the research. This was done by holding a water resources planning workshop with water users from ward 1, which falls in the Limpopo river basin in Zimbabwe[52]. The workshop was held on Friday 22 May 2009. Ward 1 is made up of six villages, and from each village the village head and three other persons (of which at least one had to be female) were invited to the workshop. The ward councillor was also invited and attended the workshop. All known water user groups,

---

[50]Bulawayo is physically located in the Gwayi catchment, which is adjacent to the Mzingwane catchment. However, through inter-basin water transfers Bulawayo gets most of its water from the Mzingwane catchment. For that reason the city is considered a stakeholder in the Mzingwane catchment and is therefore represented on the Mzingwane catchment council by its Director of Engineering Services.

[51] In this chapter the local level is taken to be the ward level. A ward is the lowest planning level in Zimbabwe's administrative hierarchy. See chapters 2 and 4 for more details on the ward as an administrative structure in Zimbabwe.

[52] See Chapters 2 and 4 for a detailed description and location of the Insiza District in which the Ward 1 is located, and of the villages in the ward. The same chapters also describe the socio-economic and physical characteristics of the Mzingwane catchment, and the Upper Mzingwane subcatchment in which Ward 1 is located.

such as irrigators from the Zhulube Irrigation Scheme, Sidingulwazi Nutritional Garden and the Smart Cooperative Garden were also invited to the workshop. Domestic water users and livestock owners were assumed to be present among the water users since domestic water use is a basic need. Gold panners, although they are water users, were not specifically invited to the workshop to avoid possible legal complications[53]. A total of 30 people (20 male) participated in the workshop. The workshop focused on mapping of water resources, identification of water needs of the people and options for water resources development.

8.4 Evolution of water resources planning in Zimbabwe

8.4.1 Water resources planning in the pre- IWRM era

The pre-IWRM water resources management period in Zimbabwe straddles the country's colonial, minority rule and majority rule period because the water law enacted prior to the country's independence in 1980 was carried into the independence era and was only reformed in 1998. Differences in access to water by different segments of society can be explained by policy differences of the governments in power during each era.

Access to water in colonial Zimbabwe was geared towards the creation of a white settler economy. The initial focus of the settlers was to find the 'second Rand'. Consequently in the early days of the colony legislation gave mining overriding rights over any other use. However, this was overturned by the 1927 Act which gave priority rights to agriculture (Vincent and Manzungu, 2004). Control of land and water resources was designed in such a way that white settlers maintained monopoly over water for irrigation (Campbell, 2003). The agricultural sector provided the political and economic base on which the white government depended. Thus water resources planning, development and allocation were based on a matrix of ideas of efficiency, modernity, white power and male supremacy (*ibid*). The government of the day developed water resources in the white areas but neglected the areas set aside for blacks.

The need for private players in water resources development and government's disinterest to arbitrate over conflicts between white water right holders led to the formation of river boards in the white commercial farming areas. The first river boards were established in the Mazowe catchment (Dougherty, 1997; Latham, 2002). Statutory Instrument 242 published in the supplementary Government Gazette of 29 August 1986 established the Middle Mazowe River Board, which was among the first river boards to be formed in the catchment. In total seven river boards were established in the Mazowe catchment. These mainly represented the interests of white commercial farmers (Moyo, 2004). They were established on a voluntary basis and were responsible for the operation of dams, raising levies on water rights, and monitoring the use of water by rights holders (Latham, 2002). The catchment hydrologist for the Mazowe suggested that river boards tended to be formed along river systems where use of water (for irrigation) was high and

---

[53] At the time of the research gold panning had been declared a criminal activity by the government (see Chapter 5).

the potential for conflicts over allocations[54] also high. The 1976 Water Act, which was in use at the time, stated that river boards must make river basin outline plans. However, in practice river boards were mainly concerned with finding and developing dam sites for syndicates of white commercial farmers. Thus the river boards were in essence more about dam development and water allocation plans than anything else. Data was generated from gauging devices in rivers and dams (Moyo, 2004). Black smallholder farmers did not participate in the planning process because they did not have access to stored water which was allocated using the plans.

8.4.2 Catchment planning in the IWRM era

In accordance with the Water Act (1998) catchment councils, together with the national water authority, are supposed to make river basin outline plans for their respective areas. The same Act states that a catchment outline plan must indicate, among other things:

- major water uses within the river system concerned;
- water allocation among sectors;
- the maximum permissible levels of pollution within the catchment area concerned;
- the phasing of any development and the order of priorities;
- potential dam sites.

Part 2(c) of the same Act states that in preparing outline plans there should be *"regard to any relevant regional plan."* Part Two Section 12 (2.a) of the Water Act (1998) states that in preparing catchment plans, the National Water Authority and the catchment council concerned shall:

> ....consult the authorities and bodies which in their opinion are likely to be concerned with the development of the catchment area or catchment areas of the river system concerned and the utilization of its water resources.

The stipulations of the Water Act (1998) show that river basin outline plans are required to capture the status of the resource (resource assessments), possibilities of resource exploitation and similar aspects which have to do with the water as a physical resource. The same Act does provide opportunities for the improvement of rural livelihoods through, for example, the requirement that potential dam sites be identified. This requirement means that river basin plans can be used to address the need for water for productive uses. As far as the process of making the plans is concerned, the law makes it clear that water resources planning must be participatory in nature. Furthermore, by stating that outline plans must have due regard for other plans, the law indicates that river basin plans must be informed by the contents of other plans (development plans being an example). How these provisions have been translated to actual practice remains to be answered. Of interest is how river basin plans relate to regional and/or national development plans, how stakeholders are engaged in the planning process, and how river basin plans address livelihood challenges. Answers to these questions are sought by examining river basin planning in Mzingwane catchment, and to what extent experiences

---

[54] Personal Communication, 28 May 2009

from the Mazowe catchment informed the plan. However, before moving on to analyse river basin planning in the Mzingwane, Box 8.1 presents steps recommended by the Global Water Partnership (GWP) when making an IWRM and water efficiency strategy. The GWP is one of the leading proponents of IWRM.

Box 8.1 Recommended steps in the making of an IWRM and water efficiency strategy

In the aftermath of the World Summit on Sustainable Development (WSSD) in 2002, the Global Water Partnership (GWP), which is one of the major proponents of IWRM, produced a handbook for developing water efficiency strategies. The handbook was developed after recognising that, although some countries were eager to develop Water Efficiency Strategies, they were facing challenges in getting the process off the ground. The handbook was therefore developed to aid countries make water efficiency strategies. As far as the content of the strategies is concerned, GWP suggests that strategies should:
1. cover institutional, financial and technological change and promote action at all levels
2. give priority to meeting basic human needs, especially access to water for the poor.
3. address the challenge of balancing the need to restore and protect ecosystems with the needs of other water users

As far as the process of making the strategies is concerned, GWP suggests that:
a. an entry point should be chosen: An entry point can be an issue, or a geographical location. For example, a country concerned with achieving Millennium Development Goals can develop a strategy aimed a harnessing water resources management towards the attainment of the MDGs. A country can also choose to focus on a geographic area in which water problems need to be addressed urgently
b. substantive issues should be defined and priorities set: Substantive issues and priorities can be set based on, for example, challenges of poverty reduction, increasing access to water, developing water resources, meeting the needs of special groups, such as women.
c. Responsibilities should be defined and stakeholders involved in the process.
d. A knowledge base should be created: This is basically about gathering the facts needed to develop the strategy.
e. Timeframes and milestones should be set

In addition, GWP suggests that a steering team to guide the process and to mobilise support across sectors should be set up. A management team should also be put in place to manage the day-to-day processes of making the strategy. Among the measures that can be taken to encourage stakeholder participation in the process include arranging informal meetings, workshops focus group interviews and media events.

Source: *GWP Technical Committee* (2004)

8.4.3 Piloting river basin planning in Mazowe

The Mazowe catchment was chosen as the test site for water sector reforms in Zimbabwe because it already had river boards which could be used to experiment with the implementation of the proposed changes to water law and policy in the country. During the reform phase attempts at river basin planning were made. In the Upper/Middle

Mazowe, one of the issues which members of the river board struggled with was, in the wake of the proposed transformation of the water sector, what were river basin plans supposed to address? This dilemma was captured in the river board's minutes of a meeting held on 10 March 1999, which record the following:

> Should it (the plan) be in place to allocate a resource, should it be in place as a social mechanism to redistribute wealth or change the economic infrastructure? (2/02/1999).

This issue appears not to have been resolved in the meeting.

The process of making the plans remained white-dominated. Minutes of the same meeting show that a questionnaire which the subcatchment developed was sent out together with the monthly magazine of the white farmers' union, the Commercial Farmers Union (CFU). Such publications were subscribed to by commercial farmers and generally had little circulation among smallholder farmers.

Apart from the attempts at planning which were done within the structures of an existing river board, there was also another attempt to make a river basin plan in the same catchment through an inter-disciplinary technical sub-committee (Moyo, 2004). The technical subcommittee was made up of stakeholder groups and technical experts gathered from the Department of Agriculture and Extension Services (Agritex), the District Development Fund (DDF) the Department of Water Development (DWD), the Mazowe Rural District Council, the Mazowe Valley Development Association and the Nyagui River Board[55]. The technical committee was multi-disciplinary in terms of the skills of the members, and apart from hydrologists from the Department of Water Development there were also sociologists and economists. The Nyagui Subcatchment was chosen to be the test-site for Integrated Catchment Planning (ICP), which had been suggested as a planning approach during the stakeholder consultations held during the reform process. The technical committee chose to involve stakeholders in the process of making the river basin plan. It is interesting to note that the lowest water resources management tier proposed by the stakeholders in the catchment was the Water User Board, but the technical committee chose to involve the actors at the ward and the village level in the planning process. This was at variance with the Water Act which legislated the subcatchment catchment as the lowest water resources management unit. The technical committee was thus proposing a management unit lower than the recommended subcatchment, and based on administrative rather than hydrological boundaries. This was a compromise position which was designed to accommodate the smallholder farmers (organised at ward level and therefore administrative boundaries), was in line with the proposed changes (of operating within hydrological boundaries), and considered commercial farmers who managed water through river boards. There was also a felt need to integrate water resources planning with planning at the Rural District Council (RDC) level (Moyo, 2004). Theoretically this would ensure that development and resource planning were seamlessly merged. Unfortunately the technical committee was unable to make much progress because its members were not working on river basin planning on a

---

[55] Personal communication, Mr. Alexander Mhizha, 05/07/2011.

full time basis. Meetings were not held regularly which retarded progress. This contributed to the failure of the Mazowe catchment to produce a river basin plan during the pilot phase of the water sector reforms.

In the aftermath of the pilot phase there was a reversal of the participatory river basin planning approach which the pilot phase had started. The Zimbabwe National Water Authority (ZINWA), upon its establishment and subsequent taking over of planning for the Mazowe catchment, began to rely on its staff to plan for the catchment[56]. How this occurred can partly be understood with reference to the land reform programme of the 2000s. The Mazowe catchment, because of its natural attributes, such as high average rainfall, moderate temperatures and fertile soils, is a region of high agricultural potential. This made it a target for politically well-connected and powerful individuals. The Fast Track Land Reform Programme resulted in white farmers being replaced by the black elite. This had repercussions on stakeholder participation in water resources management in the catchment. The new land owners were so untouchable that water allocation in the catchment became chaotic to such an extent that some dams in the catchment were at risk of being completely drained by the farmers. The few remaining white farmers in the catchment were too concerned with holding on to their land to worry about water allocation.

8.5 River basin planning in the Mzingwane

After the passing of the Water Act (1998) all seven catchment councils proposed during the reforms were established almost at the same time, and by 2000 were in place. Unfortunately the period which surrounded the formation of the catchment councils was characterised by serious political and socio-economic challenges in Zimbabwe. On the political front the country's Fast Track Land Reform Programme, particularly the manner in which the programme was executed, and other government policies led to serious tension between the Zimbabwean government and western countries in particular. Donor funded programmes, such as the water sector reforms, were affected as the donors withdrew their funds from the country (among the states that had funded the water sector reforms were the Dutch and the Germans). Hyper-inflation, which was the major economic problem, made the Zimbabwean currency almost worthless. The government was unable to fund even the most basic social services such as health and education let alone the new water management institutions.

The socio-economic challenges such as the ones identified above made development planning ineffective. For example, within a space of three years the following national (development) plans were formulated (a) the Zimbabwe Millennium Economic Recovery Plan (2001), (b) the National Economic Recovery Plan (2003), and (c) the National Economic Development Priority Plan (2004) (Mutenheri, 2009). Some of the plans

---

[56]Minutes of the Mazowe Catchment Council meeting held on 27/08/2002 show that planning was initially headed by the water quality manager in ZINWA. Later it became the responsibility of the catchment hydrologist.

formulated by the government were discarded shortly after publication having been overtaken by events. Importantly, national development planning was reduced to a fire-fighting effort aimed at reviving the economy which was rapidly crumbling. There were attempts to (re)build the agricultural base in the wake of the impacts of the Fast Track Land Reform Programme, and to resuscitate the manufacturing industry, but without success. The social component of development, such as provision of water infrastructure for drinking water to rural areas took a backseat. This is the context in which river basin planning took place in the Mzingwane catchment.

8.5.1 Participation by catchment councillors

Catchment planning in the Mzingwane catchment was led by staff employed by the national water authority, ZINWA Mzingwane catchment branch. The planning team was made up of the catchment coordinator (a chemistry graduate), the catchment hydrologist, and the catchment hydrogeologist. Notably, while the planning team assembled in the Mzingwane was composed of natural scientists, the planning technical team which had been assembled in the Mazowe catchment during the water reforms pilot phase had been multi-disciplinary. Commenting on the composition of the team spearheading planning in the various catchments of the country, an official in the Ministry of Water and Rural Resources pointed out that catchment planning was not benefitting from an inter-disciplinary approach. The catchment coordinator pointed out that ZINWA could not afford to hire consultants and as a result he and his colleagues were chosen to lead the planning process for the catchment. According to the catchment coordinator:

> We were chosen to plan because we were already in the system. We knew what was required. (5/03/2007, Bulawayo)

There were, however, genuine attempts by the national water authority to involve catchment councillors in catchment planning[57].

Minutes of the Mzingwane Catchment Council show that between 3 July 2001 and 20 June 2006 catchment planning was discussed in at least 20 meetings. In one such meeting, held on 7 January 2002, councillors were made aware that the outline plan would enable the catchment council to issue water permits because water rights which had been issued under the old Water Act were no longer in force. Water users had to apply for new permits. This was seen by the catchment council as an opportunity to re-allocate water. While water (re-) allocation stood out as one of the issues which river basin planning had to deal with, the catchment council meetings did not seem to have ironed out issues relating to what a catchment plan should address. In one meeting, for example, subcatchment councils were advised to "...*look at (their) water resources and decide what they wanted from catchment planning*" (Minutes of 7/08/2001). The minutes do not indicate if guidance was given as to what to look at or for, or to make decisions in relation to which aspects. Hints given in the meetings were vague, such as that population

---

[57] Statutory Instrument 47 of 2000 identifies stakeholder groups which have to be included in the formation of subcatchment and catchment councils. The identified groups include different farming organisations, industry and local authorities. Stakeholder representatives from these groups become councillors when elected into subcatchment and catchment councils.

figures would be needed to plan for water resources management. To guide the councillors in terms of what a catchment plan should address, the meeting of 7 September 2001 records that copies of old catchment plans were circulated among members and the councillors were advised to use the subheadings from the plans to fill in details about their areas.

Although the councillors were presented with an opportunity to contribute to the planning process they missed the chance to influence both the process and content of catchment planning. Such an opportunity presented itself in the meeting held on the 30th of July 2003. In the meeting the process of data collection was discussed at length. ZINWA was tasked with designing a questionnaire to be used to collect data. There is no record of discussions on what data the questionnaires were to gather. What is recorded in the minutes are fears of the councillors such as that the questionnaires might be too difficult for ordinary water users to understand. Consequently in the meeting it was recommended that two types of questionnaires be developed, one for technical institutions and another one for rural communities. To make the designing of the questionnaires a participatory process it was suggested that questions should come from the members of the catchment council and the subcatchment councils. Data would be collected from RDCs, large scale water users and farmers with privately owned boreholes. RDC councillors would administer the questionnaires in their wards while members of the SCCs would also be trained how to collect data[58]. However, according to the catchment coordinator, the questionnaire was only developed about two years later by ZINWA staff. On 5 August 2003 the discussion in the meeting centred on (a) data needs and (b) data collection. Sectors which were identified for inclusion in the process included:

- agriculture (both smallholder and the commercial farming sector);
- mining (small and large scale);
- local authorities;
- industry;
- institutions such as schools; and
- neighbouring countries.

It was also highlighted that there would be a need to review development plans and other sectoral data from other line Ministries.

8.5.2 Attempts at engaging RDC councillors in catchment planning

Efforts to engage RDC councillors in the planning process were mainly through interviews with RDC staff and councillors and a questionnaire survey. RDCs were regarded as the ideal level at which to gather data because they formed the interface between government, industry, mining and agriculture and other stakeholders. The catchment coordinator said that:

---

[58] SCC councillors are elected to represent stakeholder groups while RDC councillors are elected to represent wards, so data from both groups of councillors would have complemented the process of catchment planning.

We went to the RDCs of Gwanda, Esigodini, Beitbridge, Mwenezi, Filabusi and we had meetings with the councillors. We got rolling plans which we used to see their projects, population and other things. We went to RDCs because they know how many boreholes are in their areas, where the dams are, and the development going on in their areas. RDCs have the advantage that councillors from the wards attend, and government departments also attend RDC meetings. (5/03/2007, Bulawayo)

The strategy of trying to organise meetings with RDC councillors on days when RDCs had their meetings did not always work as some councillors were not interested in ZINWA meetings, or did not have the patience to go from one meeting to another. In cases where the councillors were willing to meet the planners, such as in Gwanda, collecting data was hindered by a number of factors. RDC councillors, for example, would use the meetings to ask about ZINWA's role in water resources management and question what the national water authority was doing, or had done for local water users. Planners were therefore forced to respond to these questions instead of focusing on their data collection mission. The coordinator recalled some of the questions that were raised in the meetings as:

They (RDC councillors) ask you, "what is there to manage? What are you bringing as an institution? Ward 'X' has no water, and cattle are dying then you tell us you want to manage water?" "Management is secondary, development first." "When are you constructing dams, when are you bringing water?" (5/03/2007, Bulawayo)

The challenges also stemmed from the fact that RDCs made their own development plans, which incorporated the provision of water infrastructure. They then used either state funds or donor funds they sourced themselves to implement their own plans hence ignoring ZINWA.

8.5.3 A selection of stakeholders views on catchment planning
The participation of three representatives, namely that of the representatives of Mberengwa Rural District Council and Bulawayo City Council (BCC) in the Mzingwane Catchment Council, and that of a chief representing a rural community in the same catchment is presented here. Bulawayo was included in the research on the basis of being the largest user of water from the Mzingwane catchment. Mberengwa was chosen on the basis of being a rural local authority thus would present a contrast to the participation of Bulawayo, which is a large city by Zimbabwean standards. Chief Sobhula was interviewed about her role in river basin planning as a traditional leader.

Mberengwa RDC
When asked about participation in catchment planning, the Mberengwa RDC's Executive Officer Planning said that in 2005 some people from ZINWA had come to the RDC's offices and left 33 questionnaires. The questionnaires were to be distributed among the RDC councillors, and the councillors were to respond to the questionnaires. The questionnaire sought information on location details such as name of ward, district, subcatchment and even the grid reference of water sources. It also asked details specific to dams such as depth, size of spillway and capacity. The questionnaire also asked respondents to provide details on uses of water and on water quality. The purpose of the

whole exercise was not explained to the RDC. Worse still, the councillors who were supposed to respond to the questionnaires never met ZINWA staff. He also pointed out that getting responses from the councillors was not as easy as ZINWA assumed it would be:

> There was no capacity building, people were just told to fill the papers without having been taught anything. Some of the councillors cannot even read. Moreover the questionnaires were in English and had not been translated. Some (councillors) just returned them without filling them. (16/03/2003, Mberengwa)

Surprisingly the planners from ZINWA never returned to collect the questionnaires they had left, and more than a year later the RDC still had the questionnaires[59]. Copies of the questionnaires were shown to the researcher. Those that had been filled in had some unanswered questions, while some questionnaires had been returned blank. Some councillors were said to not have returned the questionnaires at all. Asked why he thought ZINWA wanted the data, he responded:

> "Maybe they want to make people pay." (16/03/2007, Mberengwa)

In fact he saw the whole point of planning as being just a paper exercise, reasoning that if, for example, the catchment council was planning to levy water users the central government had the power to nullify any programme which was felt to be against the people. He suggested that ZINWA should actually follow the example of the planning processes which are used by RDCs, namely involving communities (see Box 8.2).

---

[59] The draft plan which the Mzingwane Catchment produced does not show evidence that data collected through the questionnaires were used. In no other RDC was ZINWA reported to have used questionnaires to collect data.

Box 8.2 Participatory planning at RDC level.

Rural District Councils use a bottom-up approach to develop their plans. A district is made up of several wards. Below the ward are villages (see Chapter 4 for a more detailed description). Each ward has a Ward Development Committee (WADCO) which is chaired by an elected ward councillor. Members of the WADCO come from the Village Development Committees (VIDCOs). Developmental issues from the villages are presented to the VIDCO which in turn passes them on to the WADCO. The ward councillor can present these issues to the RDC, or to the Rural District Development Committee (RDDC) which incorporates them into the Rural District Development Plan (RDDP). The RDDC is made up of heads of government departments at the district level and heads of other institutions such as non-governmental organizations, and sectoral groups in the district[60]. The latter are also supposed to make submissions on behalf of their constituencies, or to enlighten the RDDC on what their plans are. This is meant to harmonise RDC plans with other sectoral plans. The RDDC has subcommittees which look at different sectors generally covering natural resource development/conservation, water, social services and infrastructure development. To prove that the submissions made by the councillors to the RDDC or the RDC have been agreed upon by the community, minutes of meetings held to come up with the projects and issues from each ward accompany the submissions.

The situation reported in Mberengwa was not very different from the participation of other rural district councils. The impression of RDC councillors and staff was that ZINWA was carrying out a survey to find out the number of boreholes and small dams in the catchment. The perception was that this data would be used to levy water users. This was because ZINWA was not only asking about the number of boreholes, but also about livestock numbers and numbers of irrigators in each ward. However, the execution of the data collection process made the efforts come to nought. Councillors who were supposed to help with the data collection were not informed of the process, and there were suspicions that the national water authority would use the data to add to the financial burdens on smallholder farmers through a new levy.

Bulawayo City Council
Bulawayo City Council (BCC) was represented on the Mzingwane Catchment Council by its Director of Engineering Services, Mr. Sobhula. He holds a post-graduate qualification in engineering, and his entire professional career has been in the water supply and sanitation services sector. He has worked for an engineering company consulting in water, and has also worked with different municipalities as an engineer. He joined the Bulawayo City Council as a Senior Engineer in 1988, and over the years rose through the ranks to eventually become the Director of Engineering Services in 2002. He describes himself as someone with a passion for water. He describes the level of discussion in the catchment council as 'mediocre,' and the meetings themselves as 'mundane.' He feels too much time is spent discussing permits and finances instead of more serious issues. In his opinion catchment meetings should be about planning for water resources, where to site boreholes, dams and so forth.

---

[60] Organisations which exist in each district tend to vary depending on the level of economic development and also the natural resource base. In Insiza District, for example, there are organisations which represent miners (Small Scale Miners Association being one of them).

Mr. Sobhula said that it was vital for BCC to participate in catchment planning since Bulawayo is the largest user of water from the Mzingwane catchment. The city of Bulawayo is physically located in the Gwayi catchment but gets most of its water from the Mzingwane catchment through inter-basin transfers. Water is crucial to the city because of the semi-arid conditions which characterise the region in which it is located. The importance he attaches to participation in water resources management is evident in that he attends meetings of both the Mzingwane and the Gwayi catchments. He said that he never missed catchment council meetings except when he was out of the city or had very important reasons for not attending. His participation is made easy by that the two catchments have their council meetings on separate days. Furthermore, Gwayi meetings are held within the city so he can walk to the venue which is not far from his office. When the Mzingwane Catchment Council has a meeting he usually travels together with ZINWA staff on ZINWA's vehicle. Indeed throughout the period of this research he attended the meetings which the Mzingwane Catchment Council organized.

The main interest of the city was its sources of water supply. He said that he presented the findings of several studies which had been done on the water supply for the City of Bulawayo, including studies on dams such as Glass Block Dam, Lower Tuli Dam and Mtchabezi Dam. He also said that he had enlightened ZINWA on the population and economic projections for the city so that these could be used in decision making by ZINWA.

> The city is big and a big consumer of water. Its participation in catchment planning is complicated. ZINWA is the main planner and it tells the city where the water is and the city treats the water and supplies it to the residents. When the city is about to reach the limits for its water supplies it informs ZINWA. The city makes projections on its water needs based on its population figures and it passes the information to ZINWA. (13/03/2007, Bulawayo)

He pointed out that it was up to the catchment council to use the data he had presented, but his role was to ensure that the city's position was known to the catchment council. It is clear that Mr. Sobhula adopts a proactive approach with respect to catchment planning.

Traditional leaders

One of the members of the Mzingwane Catchment Council is a traditional chief who represents traditional leaders and a rural constituency. Chief Pathe is one of the few female chiefs in the country. She is chief (*inkosi*) in the Gungwe area of Gwanda. The area is in the Mzingwane's Shashe subcatchment. As a traditional leader she has a seat on the Gwanda Rural District Council and the Mzingwane Catchment Council. However, she says that she does not always attend catchment council meetings because she cannot afford the transport costs. The costs of using her own car to travel for catchment council meetings are too high. Although she holds an important position in the community and is concerned about the lack of development in the area she was not aware that ZINWA was making a catchment plan and therefore she did not participate in the process.

8.5.4 Outcomes of river basin planning in the Mzingwane

The river basin planning process in the Mzingwane catchment resulted in a draft plan that was produced in 2006. The draft plan is a 69-page document, divided into 7 chapters. It has a 20 year planning horizon, covering the period from 2005 to 2025. The point of departure for the plan is the water-related hardships which rural water users in the catchment face, and that is contained in the plan's Chapter 1. Table 8.1 gives the structure of the draft Mzingwane catchment outline plan.

Table 8.1 Table of contents of the draft Mzingwane catchment outline plan

Chapter	Contents
Chapter 1	Introduction and objectives of the plan
Chapter 2	Geographical extent of the catchment and main socio-economic activities in the catchment
Chapter 3	Catchment hydrology and the hydro-geological factors which influence groundwater resources
Chapter 4	Water demand and use
Chapter 5	Water Balance
Chapter 6	Water resources development projects
Chapter 7	Catchment development scenarios, priority of water use

Although the river basin plan which the Mzingwane catchment made adheres to the stipulations of the Water Act (1998), the following omissions are highlighted here for their importance to river basin management in general and to socio-economic development in particular:

1. The catchment is not placed within the Limpopo basin and therefore the transboundary dimension is ignored. Thus there appears to be no consideration that developments in the Mzingwane catchment may impact on riparian users located downstream of the catchment
2. The plan fails to incorporate into its analysis the City of Bulawayo, one of the largest water users in the catchment although the city is physically located in the Gwayi Catchment
3. Land use planning is only mentioned in relation to the development of irrigation schemes
4. The plan is silent on green water
5. The plan does not address challenges of access to water at the local level

The process of the making of the Mzingwane catchment outline plan was largely dominated by the Zimbabwe National Water Authority (ZINWA), Mzingwane Branch, and in a way was a top-down process. This can be explained by that the process started within a year of the formation of catchment councils, thus the newly formed organisations were still not fully functional and one can even say, also not yet fully conversant with water resources management issues in the catchment. The catchment councils had been rapidly formed so much that some stakeholder groups were not aware of their existence. Lack of capacity in the catchment council contrasted that within

ZINWA which inherited staff from the Department of Water Development (DWD). Thus even though ZINWA tried to involve the councillors in the process of making the catchment plan, the national water authority in practice dominated the process. The Catchment Hydrologist, Hydrogeologist and Coordinator were the main actors involved in the planning process, and they designed the data collection instruments and did the actual data collection. The data collection process involved the use of questionnaires (see Appendix 1), key informant interviews and secondary data. Although the initial plan of the planners was to design different questionnaires to be administered to different stakeholder groups, in the end only one questionnaire was used, and it focused on the identification of different water resources. Some of the questionnaires were handed to Rural District Councils (RDCs) and RDC councillors were to respond. However, in some cases there were no attempts to translate the questions which were in English, something which could have helped in making responding to the questionnaires easy in rural communities. In Mberengwa it was reported that the ZINWA did not collect the questionnaires it had left with the RDC. The fact that the national water authority failed to collect the questionnaires must be understood in the context of the economic environment in which planning took place. At the time the catchment councils were not financially stable and relied on public funds. However, because of the economic crisis which the country was facing there were inadequate funds allocated to run even the most basic of services, such as health, which meant that 'non-essential' services, such as river basin planning, fared even worse. The 'un-popularity' of ZINWA in the country also affected data collection. Planning took place at a time when the national water authority had been handed over water services of cities such as Harare, and the government was fighting to take control of the same services in Bulawayo. In places where ZINWA had taken over water services there were allegations of mismanagement and rising cost of water. Water users therefore perceived the data collection process to be either a step towards the takeover of water services, or a way of charging water users. Thus in the end the process relied on desk-top studies and other outdated data which ZINWA already had.

The next section moves on to analyse the making of a ward water plan. The ward water plan was made as a way of contrasting planning done under the guidance of legislation and as done by employees of the national water authority to planning done by water users who plan guided by their livelihood conditions.

8.6 The making of Ward 1 Water Plan

As part of this research the water users in Ward 1, Insiza district (which falls within the Mzingwane catchment) developed a ward water plan. This was done upon the initiative of the researcher, who also facilitated the process.

In order to establish the adequacy of available water infrastructure and sources of water in the ward, participants at the ward workshop were put into groups according to their villages[61]. They were asked to draw maps showing the location of all the water sources in

---

[61] See section 8.3 for a description of how the ward workshop was convened and on the participants who took part.

the village. For each source of water participants were asked to identify uses to which the water was put, the issues around each source, for example, for boreholes they were to state whether the borehole was working or not, and the institutions managing that particular source of water.

Participants were asked to discuss what a water resources plan is, and what it should contain. Some of the reasons which were given for the need for a plan included:

- to identify possible sources of water;
- to address livelihood problems related to water resources, for example, diseases;
- to deal with problems caused by gold panning; and
- to develop irrigation schemes and gardens and get good harvests.

The main categories around which the ward water plan could be developed were given as:

1. Water resources development and access to water;
2. Institutional development/capacity building;
3. Watershed conservation management; and
4. Maintenance of water infrastructure and training of pump minders.

Participants were asked to form groups which were made up of people from the different villages in the ward. This was done to ensure that the plan would not be limited in scope to one village but would consider the ward as a whole. Each group used the subheadings which had been agreed upon to make their plan. In all cases the ward water plans which were made were very specific in terms of what the water users wanted addressed. Outputs from the different groups were then discussed and used to compile one plan. The following presents the consolidated outcome of the group meetings.

Addressing the challenge of access to water
The problem which was identified here was that access to water for both domestic and productive uses was poor across all villages. This was attributed to the fact that in some villages the geological conditions made it difficult for the water users to dig wells. A common problem across all the villages was the breakdown of water infrastructure. Increased access to water was presented as a means for enabling communities to engage in agricultural production. All groups listed the number of boreholes, small dams and wells they wanted to be dug in the ward, and in the end the participants agreed on the need for the following:

- Boreholes:[62]   at least 30 per village
- Deep wells: 10 per village
- Small dams: 12 in the whole ward

---

[62] In Chapter 4 reference was made to that in ward 1 shallow wells which have been fitted with pumps are commonly referred to as boreholes. However, in this case water users wanted developmental agencies to sink boreholes for them, so specifically here a borehole refers to a hole of about 20 cm in diameter which is drilled 50m deep or more into the ground, and from which water is pumped out using a hand pump.

It was claimed that these figures were arrived at by taking into account the population of each village. However, the suggested number of boreholes per village appeared very large, especially in relation to size of village by population and spatial area. There was also no consideration of the nature of the underlying rock as a determining factor in the siting of water infrastructure. However, the importance of this exercise was that it enabled the water users to link the activity of identifying water resources and their uses to the needs of the households in the ward. Thus the participants were in a way using data on the available water resources to suggest how their water situation and livelihoods could be improved, which meant that in essence they were not making just a resource plan.

Planning for institutional aspects of water resources management

The major point that came from the discussions was that Water Point Committees (WPCs) were not effective in their work. This was said to be caused by a number of factors, such as that these committees lacked authority to enforce rules. It was also pointed out that committees were unable to facilitate the repairing of broken down boreholes. The committees were also said to be unable to procure spare parts. This was attributed to the fact that water users were not making the financial contributions needed to meet the costs of repairs and other expenses. The need to strengthen water point committees through training was suggested as a possible solution to the problem. The training that was being referred to related to aspects such as record keeping, financial management, minute taking and other areas on the running of committees at community level. It was suggested that there was need to bring village heads into the management of water resources to give authority to the committees, and that village heads should enforce rules and encourage water users to make contributions. It was also suggested that there was need to train pump minders so that repairs and maintenance of infrastructure could be carried out at the village level.

Watershed management/catchment protection

On this point the participants showed concern for the effects of gold panning on the environment. Erosion, the reduction of channel size and the siltation of small dams were identified as the main issues. However, no clear solutions were suggested. This was probably because the water users felt the problem could not be solved without impacting on their own livelihoods (compare with Chapter 5).

8.7 Discussion and conclusions

This chapter addressed questions which revolve around the process, context and outcomes of river basin planning. However, it is important to also analyse the relationship between resources planning and development planning. It appears that at both the global level and the national level resource planning has historically supported socio-economic development. The TVA, for example, had a strong development-focused approach. The huge amounts of data amassed by the agency fed directly into infrastructure development projects. This approach suited the context in which the TVA was formed, namely the Great Depression . Similar trends are observable in colonial

Zimbabwe when the government of the day, working together with and for white farmers, built the national economy and a particular social structure through water resources planning and development. In fact it can be argued that the whole colonial machinery associated with water resources management spent most of its energy on water resources development and water allocation - all for the purpose of creating a white dominated state. However, and this is the main finding of the chapter, the current water resources management paradigm appears to have dropped the development aspect, to the detriment of livelihoods. However, this needs to be put into context.

The Mzingwane river basin plan was made at the peak of the socio-economic crisis which gripped Zimbabwe in the 2000s. At the time inflation was a huge problem. Central government could hardly fund even the most basic services, such as health and education. The Mzingwane catchment council was directly affected by the unavailability of funds. In the same period that the Mzingwane catchment council was making its river basin plan central government crafted several national development plans. Some of the development plans were discarded hardly a year after being drafted and publicised. Constrained development planning by central government cascaded down to affect river basin planning.

The socio-economic context clearly affected the process of river basin planning, particularly the engagement of stakeholders. As was shown in the chapter, the manner in which stakeholders were engaged raises serious doubts over the sincerity of the planners. Meetings through which data collection was supposed to be done appear not have been planned for effective participation. In some cases data collection instruments were not collected from the communities.

The Water Act (1998) requires river basin plans to take into account regional and any other relevant plans. This, at least in theory, enables resource plans to inform and be informed by development plans. As far as livelihoods are concerned development plans made at the district level by Rural District Councils (RDCs) are among the most relevant to smallholder farmers. The district level is critical because it brings together different stakeholders. Most government departments have representation at the district level and the rural populace is also represented at the same level through ward councillors (Van der Zaag, 2005). Government programmes and projects also tend to be implemented at the district level. In short, the district level interfaces with the local level (which is important for the identification of developmental needs) and the government (which funds developmental projects).These reasons make it crucial that river basin plans take into account district development plans, especially for purposes of addressing livelihood needs. The provision of the Water Act (1998) which compels catchment councils to take into account other plans when making river basin plans is therefore a positive feature. However, the reality is that the process of getting different plans (in this case, the river basin plans and district plans) to 'speak' to each other is confounding. One challenge is the issue of the scale at which the different plans are made. River basin plans are made at the catchment level. Without exception, each catchment contains several districts. There are seven districts in the Mzingwane catchment. Within the catchment are also a number of urban centres which make their own development plans. Collating the different plans into one coherent unit can therefore be a mammoth task. The conclusion is that on paper the provisions for different plans to feed into each other are there. However, what needs to be improved is the strategy for operationalizing the provisions.

The content of river basin planning in Zimbabwe is guided by the Water Act (1998), but what bearing does that have on the improvement of livelihoods? The contents of the river basin plan of the Mzingwane catchment shows conformity to the stipulations of the Water Act (1998). It can be argued that although the plan, judging by the nature of its content, is basically a resource plan, it does indeed contain data and information useful for socio-economic development, albeit indirectly. Development planning has to be informed by hydrological facts, which is what the river basin plan does provide. An important aspect of the plan is that it actually identifies potential sites for dam development and makes an analysis of available water resources. Such information is the meat on which development planning depends. This therefore answers the question whether resource plans made in the context of IWRM can contribute towards development planning. A shortcoming of the plan is that it does not articulate how the water resources facts (on hydrology, for example) it gathered can be translated into actual projects which can contribute towards the improvement of livelihoods.

Although the plan does meet some of the stipulations of the Water Act (1998), it appears to fall short in a number of aspects when measured against the recommendations of the GWP. As far as the content is concerned, the plan, for example, appears not to tackle how the water needs of the poor can be met. Furthermore, it also does not identify institutional nor technological changes needed to improve water resources management. The GWP (2004) suggests that these aspects of water resources management should be prioritised and be included in the plans. However, the fact that the process of the making of the Mzingwane river basin plan was firstly presented to the Mzingwane catchment council suggests efforts at creating a steering committee, which is what the GWP recommends. As was shown in the thesis, the councillors attempted to identify issues and set priorities, all which is recommended by the GWP. However, although these aspects of planning were dealt with in the meetings, they appear not have been considered when the actual plan was made. Thus the draft plan which was produced does not clearly identify issues it seeks to tackle. GWP recommends that a plan should clearly identify issues or a geographic locality it seeks to deal with. Arguably, this lack of a defined agenda resulted in the plan neither dealing with access to water by the poor, nor water resources development issues in a concrete manner. This can be said to be the major weakness of the plan when measured against the recommendations of GWP.

The transboundary dimension of water resources management appears to be missing in river basin planning. The Mzingwane catchment plan is limited in scope to the portion of the Limpopo river basin which lies within Zimbabwe. From a legal perspective, rightly so because the mandate for the catchment council and the national water authority to carry out river basin planning is derived from national legislation. However, developments within the catchment have downstream implications. These downstream-upstream links explain the formation of LIMCOM, which was formed to coordinate developments within the Limpopo river basin. It is therefore surprising that the river basin plan which the Mzingwane catchment council produced takes a narrow view of water resources, not considering potential downstream impacts. One reason for this narrow perspective could be that river basin organisations are not represented in the basin commission (Fatch *et al.*, 2010).

It may be added that however planning is conceptualised and executed, in the final analysis it is influenced by political choices made, the processes followed, and the objectives set. In the context of southern Africa, Zimbabwe included, there is still need for a development-oriented approach to water resources management. This approach should be informed by river basin plans. There is therefore need for political choices which support concrete measures to be taken to meet the water needs of rural water users. This calls for a policy shift, a return to resource planning aimed at development.

Chapter 9

Conclusions and recommendations

9.1 Introduction

This thesis took a bottom-up approach to analyse the drivers influencing the water resources management-livelihoods nexus. Analysis was made at the local level, which for purposes of this study was taken to be the level at which water is abstracted and used. Undoubtedly water resources management covers a wide spectrum, but this study purposefully targeted water resources management as it occurs: in the management of water sources and water infrastructure, in catchment management, as field level water management, as urban water services management, and in river basin planning. These different arenas were chosen because they cover a fairly wide spectrum of the livelihoods of local actors. The findings made, and conclusions reached by this study are presented next.

9.2 Findings of the study

This section presents findings and conclusions of the study, starting with a focus on local actors and issues in water resources management and ending with actors and issues at the river basin level.

9.2.1 Water resource management at the local level

Chapter 4, which was the first of the empirical chapters, tried to answer the question, *what drives practices in water resources managed at the local level ?* To answer this question, practices of water users were analysed at different sources of water and water infrastructure, specifically a borehole, a wetland and a wind-powered water infrastructure. A major finding of the chapter was that practices in water resources management at the local level are occurring outside the IWRM framework but respond to local socio-economic and physical factors. The adoption of IWRM and the subsequent formation of catchment and subcatchment council in a way represent an attempt to influence local practices in water resources management. However, the absence of IWRM-based institutions at the local level seems to nullify assumptions made at the international level, such as that institutions can be parachuted to the local level. The reality is that water users try and solve challenges relating to access to water and infrastructure maintenance using locally available resources, institutional or otherwise.

Evidence suggests that the assumption that the solution to water resources management problems is a well-structured framework with a clearly defined institutional mandate is flawed. Implementing the IWRM framework has resulted in boundaries for managing water resources being set at the catchment level, and stakeholder platforms being established. However, this neat framework is confounded by that in reality, at the local level water resources management follows politically constructed boundaries. Furthermore, socially constructed relations among water users have a strong bearing on how water resources are managed. In this particular case perceptions of impending water scarcity, and the contested ownership of waterpoints, were found to be very influential in shaping practices of water users. At other waterpoints more harmonious social relations existed and a sense of moral obligation among water users to cooperate with each other.

The study also found that at waterpoints where water was used for productive uses, infrastructure tended to be better maintained than where it was used only for domestic uses. In the light of this finding improving water resources management could therefore be tied to improving livelihoods through increasing access to water for productive uses. This appears to be recognised by proponents of IWRM, such as GWP (2004). GWP suggests that IWRM and water efficiency strategies should aim at tackling poverty and increasing access to water by vulnerable groups in society, such as the women and the poor. The challenge is, how can this recommendation be implemented?

It appears that to date the rational choice theory (Ostrom, 1990) has informed possible solutions to the challenge of improving water resources management. The gist of the theory is that institutions which meet specific conditions can be purposefully crafted. However, the findings of this study differ. It was found that water resources management institutions which were linked to other socio-economic spheres functioned better than those that were specifically created to manage water resources alone. This seems to confirm Cleaver's (2000) observations that in practice water users transfer institutional arrangements from one context to another. She refers to this malleability of institutions as institutional bricolage. Importantly, institutional bricolage acknowledges that resource users respond to local contexts as they manage resources. In view of this finding, one can conclude that rather than forming new institutions at the local level, efforts should be directed at coupling the management of waterpoints to the management of other resources, or to the organisation of other activities already taking place in the community. One can also add that, as was shown in the chapter, water resources management can be better understood by applying a theoretical lens that makes it possible to analyse both the social and physical/material factors. Local context, which Cleaver (2000) shows to be influential in water resources management, consists of both the social and the physical. This chapter showed that water use practices and local institutions are influenced by both factors. It can therefore be argued that the tragedy of IWRM could be that, while it is driven by an understanding of the global state of water resources, it is not informed by an adequate understanding of social factors relevant to the different local contexts.

9.2.2 Environmental management

The focus in Chapter 5 was on understanding how local actors try to 'sustain' livelihoods and the environment (catchment management). The context in which the analysis was made was characterised by an adverse socio-economic and physical environment. The chapter showed that actors at the local level exploit the physical and the institutional resources at their disposal to sustain themselves. This results in seemingly contradictory strategies. On the one hand actors engage in environmental reclamation projects, and on the other hand damage the environment through gold panning. The environmental reclamation project was implemented by non-state actors as a Food For Work (FFW) project. Such projects are perceived as being able to provide the poor with a safety net while investing in the maintenance of valuable public goods (Barret *et al.*, 2003). This seems to make it possible for environmental management to be carried out in tandem with livelihoods sustenance. The adverse socio-economic context in Zimbabwe, which resulted in shortage of goods and services, made local actors receptive to FFW projects. However, the chapter showed that when non-state actors stopped handing out food items, the reclamation project collapsed. This supports the view that although FFW projects try to address livelihood challenges, the reality is that the majority of the world's vulnerable population are affected by chronic hunger (*ibid*) which cannot be solved through piecemeal and short-term efforts. The chapter also found that catchment management projects were being championed by non-state actors, not river basin organisations. This raises the question of how institutional roles are defined in the IWRM framework, and in practice? An important finding of the chapter was that the poor are willing to participate in catchment management projects. In India where similar projects have been undertaken it has been shown that broad based participation contributes towards the success of projects (Vishnudas *et al.*, 2008). However, they suggest that sustainability of such projects is linked to, among other things, how projects address socio-economic issues affecting smallholder farmers.

Interestingly, within the IWRM framework the importance of economic issues is recognised. The fourth Dublin Principle states that water has an economic value and should be recognised as an economic good. The question in need of an answer is, how can that Principle be made to benefit smallholder farmers affected by poverty? Poverty influences the actor-environment relationship. Treating water as an economic good which should be allocated to the 'highest bidder' can potentially disadvantage smallholder farmers and worsen environmental management. A reformulation of the fourth Dublin Principle which makes economic considerations in water resources management benefit the poor is therefore needed. If economic benefits accrue to the poor environmental management can potentially become sustainable. Although the concept of water control considers socio-economic factors in the analysis of water resources management, the concept has mainly been used to analyse socio-political factors at the expense of economic factors. This calls for the need for analytic tools which can probe deeper into how economic factors influence water resources management at the local level.

9.2.3 Conservation agriculture

The management of agricultural water has mainly focused on water allocations and use efficiencies in irrigation, yet the majority of smallholder farmers in southern Africa depend on rainfed agriculture. In Chapter 6 narratives of different actors were used to analyse how field level water management techniques are being promoted among smallholder farmers, and how these are influencing dynamics among different actors. Narratives were used as a way of accessing social reality in smallholder farming (Hyden, 2008). The chapter found that conservation farming was being promoted among smallholder farmers mainly by non-state actors using agricultural inputs as incentives to encourage the adoption of conservation agriculture. A declining economy had the twin effects of incapacitating the civil service, and making agricultural inputs scarce and expensive. Input incentives, such as seeds, made smallholder farmers receptive to practices of conservation agriculture. Planting basin tillage, mulching, and micro-dosing of fertilizers were found to be the main aspects of conservation agriculture being promoted. Planting basins in particular were being promoted on the argument that they capture water and therefore increase soil water available to crops. However, the chapter found that once incentives were removed smallholder farmers lost interest in conservation agriculture. The strategic adoption of conservation agriculture by smallholder farmers seems to support the views of Baudron *et al.* (2011) that in reality conservation agriculture itself goes against the practices of farmers in this category. This begs the question, how can (high) agricultural productivity in Africa's smallholder farming sector be sustained?

To answer the above question it is important to draw lessons from the 'green revolution,' which preceded conservation agriculture. The 'green revolution' that occurred between the 1960s and 1990s in most of Asia and Latin America (IFPRI, 2002) could not be sustained in Africa. An important factor explaining this variance was that the green revolution targeted crops that were not widely cultivated in Africa, such as rice (Asiema, 1994). Furthermore, the agro-ecological conditions of Africa were not similar to those of Asia, yet the formula that had worked in Asia was supposed to work on the continent (*ibid*). It can be argued that the failure of the green revolution in Africa was a result of limited understanding of the realities on the continent, these realities being the agro-ecological conditions and the livelihoods of smallholder farmers. This suggests that research must not only seek to understand smallholder agronomic practices, but must also try to understand the livelihoods of such farmers. Van der Zaag (2010), for example, suggests that there is need for the creation of knowledge on local variations in water and soil nutrient availability. This could potentially contribute towards improved smallholder farmer productivity.

9.2.4 Struggle for the control of Bulawayo water services

In Chapter 7 the endeavour was on trying to understand how context accounts for contestations over the control of urban water services. The case was based on the proposed takeover of the city of Bulawayo's water services. The chapter showed that the proposed takeover was premised on reasons of efficiency, and those opposing the takeover cited reasons of legality. However, it was found that the symbolic meaning of water plays a major role in influencing water resources management. This was found to be particularly true in the context of an adverse socio-economic and political environment. The government wanted to capture the city's water sector ostensibly because of its symbolic meaning to the residents. The city's water services represented defiance, to nature as the city is in a semi-arid environment, and to central government as the city had built its own dams over the years. Water services represented the will of the residents to make the city thrive under the most adverse conditions. This explains to a large extent why water coalitions against the government emerged. To the government capturing the water services would mean that it would exert direct control over an urban authority. This would represent a triumph over urban areas that were increasingly becoming opposition party strongholds. Thus, contrary to prescriptions of international financial institutions, economic efficiency alone cannot explain why certain water resources management models are accepted while others are rejected. In some cases models are accepted or rejected based on the social value attached to water. When one applies the findings of the chapter to IWRM, in particular how the framework has been promoted, one can ask what role have perceptions about the framework played in its adoption or rejection? As Heynen et al., (2006) note, water management frameworks being promoted by the World Bank and similar bodies are inclined to neo-liberalistic ideology. They consider the promotion of good governance in the water sector as being linked to citizen (dis)empowerment. One can extrapolate the findings of this chapter to suggest that within the developing world, half-hearted attempts at implementing IWRM are in fact linked to governance issues in general rather than to water-sector specific factors.

9.2.5 River basin planning

Chapter 8 tried to account for processes in river basin planning, and whether their outcomes match the livelihood realities of water users. From the onset it needs to be highlighted that river basin planning provided an opportunity to take a bird-eye's view on water resources management in the Mzingwane catchment. Thus it is imperative that this discussion relates to all the empirical chapters presented in the thesis.

As was discussed in Chapter 2, Zimbabwe faced immense socio-economic challenges which roughly started at the turn of the millennium. Among other things, the crisis severely affected the economy, and that in turn affected government-funded activities and programmes. The period coincided with the making of the Mzingwane catchment plan. Given the challenges that were prevailing in the country, it is actually remarkable that the Mzingwane catchment council was able to produce a plan at all. The chapter found that, as far as the Water Act (1998) is concerned, there are opportunities for participatory planning and for making river basin plans inform and interact with other sectoral plans. This can potentially enable river basin planning to contribute towards socio-economic development. Key clauses in the legislation include the requirement that river basin plans consider regional plans, and that relevant stakeholders be consulted. However, the actual making of river basin plans was found to be a challenge. For example, the collapse of the economy made participatory planning extremely difficult as the process was under-funded.

Findings of Chapter 4 showed that a major challenge in the catchment is that of access to water at the local level. This need for improved access to water at the local level was further highlighted in Chapter 8 by the ward plan which the water users made. It is therefore surprising that the draft plan which the Mzingwane catchment council made appears not to address such challenges. The plan, for example, makes no reference to how domestic water needs in the catchment are to be met. Chapter 5 showed the challenges of sustaining environmental resources management while at the same time meeting livelihood needs. These challenges were shown to be missing in the draft catchment plan. In Chapter 6 the focus turned to water resources management at the field level, showing the challenges posed by rainfall variability in the catchment. Sadly, river basin planning also appears to miss the link between water resources management and smallholder farmer productivity. River basin planning, as was analysed in the chapter, demonstrates that although statutory prescriptions do exist to guide processes in water resources management, these may come to naught upon encountering reality. This reality includes the economic environment which determines the capacity of river basin organisations to fund water management processes. The realisation that most developing countries do not have the capacity to implement 'full IWRM' have resulted in calls being made for countries to implement 'Light IWRM' (Moriarty *et al.*, 2010). 'Light IWRM' is considered to be more practical, incremental in approach and focused on the needs of the water users. However, the fact that there are calls that IWRM should be altered can be taken as evidence that the framework is not geared to meet local contexts in the developing world.

9.3 Concluding thoughts

In this study a multi-foci approach was taken to analyse the drivers of the water resources management-livelihoods nexus. Specifically, the management of water for domestic and productive uses, the catchment management-livelihood strategies nexus, field level water resources management, urban water services management, and river basin planning were studied. Such an approach is rare in literature. Furthermore, in many cases analysis tends to take a top-down approach, which often results in livelihoods either not receiving adequate attention or being omitted completely. This study therefore demonstrated the importance of an approach which takes a bottom-up view of water resources management and livelihoods. Such an approach is particularly important where policy informed by real livelihood challenges is to be crafted. A major finding of the study was that of the importance of local context as a driver of practices in water resources management. The adoption of IWRM and other water management frameworks which preceded it gives the impression that global discourse and international actors are the major drivers of practices in water resources management. This explains why water management frameworks end up being parachuted into communities. However, the reality is that communities are enmeshed in a socio-politico-economic and physical environment which exert influence on local actors on a day-to-day basis. As actors react and respond to the force exerted by the socio-politico-economic and physical environment, water resources management and livelihoods are shaped in a particular manner. Actors, in other words, do what they reason they have to do to 'secure' their livelihoods. Where it is strategic to their livelihoods, certain discourses and practices are adopted. This can include those practices that are considered by global actors, such as NGOs, to be environmentally sustainable. In the context of this thesis, catchment management and conservation agriculture were clear examples.

Although the main objective of IWRM is to 'integrate', in practice this has not been easy to achieve, from both a conceptual and an operational perspective. At the conceptual level it has become evident that the management of blue water, for example, takes centre stage in the water sector while green water remains largely ignored. This is a critical omission given that the majority of water users in southern Africa derive their livelihoods from a dependence on green water. Furthermore, this study showed that river basin organisations appear not to be taking centre stage in catchment management initiatives.

This too is puzzling given the relationship between the state of the catchment and water resources. As was shown in the thesis, water resources are threatened by physical processes taking place in the catchment. The thesis showed that actors spearheading catchment rehabilitation were not from river basin organisations. The government's response to catchment degradation was not from a water resources management perspective. This is understandable given that policy prescriptions came from an entirely different sector, specifically the economic sector. Institutional roles therefore need to be clearly defined, and where catchment management is concerned, the water sector should champion policy development. At the operational level one of the major challenges is that of merging different institutions. In this thesis the problem of institutional (mis)fit was shown to be rampant. This problem appears to have been multiplied by that challenges in water resources management are often construed as being rooted in institutions. This explains the formation of waterpoint committees in the late 1980s, and the formation of catchment and subcatchment councils in the late 1990s. In reality this has resulted in institutional pile-up. One can argue that to date it appears that the most common solution to dysfunctional institutions has been to create new ones. As was revealed by this study, there is evidence that some institutional challenges are symptoms of deep seated livelihood problems. In this study, for example, where water resources were used productively institutions tended to be functioning better than where water was used only for domestic purposes. Policy therefore must seek to attack poverty and other livelihood challenges.

Given the challenges which Zimbabwe has faced in the recent past it is commendable that significant steps have been made towards implementing IWRM. Catchment and subcatchment councils have been set up in all the country's seven catchments. The fact that efforts are being made to make catchment plans signifies life within the catchment councils. To this end it can be concluded that the country has embraced IWRM and is trying to flow with current approaches in water resources management. However, findings of this study seem to suggest that one shortcoming is that the country has not tried to make adjustments needed to make IWRM suit the local context. Poverty, access to water and environmental degradation are among the major challenges which the country faces. Improved water resources management can contribute towards addressing these challenges, yet indications are that these issues are not on top of the water resources management agenda in the country. Arguably, Zimbabwe, and other developing countries, need to conceptualise and implement IWRM in a manner which reflects their socio-economic and environmental needs. Furthermore, given the demonstrated influence of local level drivers on water resources management and livelihoods, it can be argued that the challenges in water resources management cannot be solved outside of the wider socio-politico-economic realm. Interventions implemented in the developing countries must therefore try to resolve more than just the challenges that manifest in the water sector.

References

Adams, W.A. 1992. *Wasting the rains: rivers, people and planning in Africa.* Earthscan Publications Ltd, London.

Alexander, J. 2006. *The unsettled land: state-making and the politics of land in Zimbabwe 1893-2003.* James Currey, Oxford.

Allan, T. 2003. *IWRM/IWRAM: a new sanctioned discourse? Occasional* Paper 50. SOAS Water Issues Study Group. King's College London, University of London.

Allenby, B.R. 2000. Environmental security: concept and implementation. *International Political Science Review* 21(1). 5-21

Andersson, J.A. 2007. How much property did rights matter? Understanding food insecurity in Zimbabwe: a critique of Richardson. *African Affairs* 106(425). 681-690.

Andersson, J.A. and Giller, K.E. 2012. On heretics and God's blanket salesmen: contested claims for conservation agriculture and the politics of its promotion in African smallholder farming. In Sumberg, J. and Thompson, J. (eds). *Contested agronomy: agricultural research in a changing world.* London. Earthscan.

Asiema, J. 1994. Africa's Green Revolution. *Biotechnology and Development Monitor* 19. 17-18.

Barnett, J. 2001. *The meaning of environmental security: ecological politics and policy in the new security era.* Zed Books, London.

Barrett, C.B., Holden, S., and Clay, D.C. 2002. Can Food-for-Work Programmes Reduce Vulnerability? World Institute for Development Economics Research (WIDER). Discussion Paper No. 2002/24.

Barron, J., Rockström, J., Gichuki, F., and Hatibu, N. 2003. Dry spell analysis and maize yields for two semi-arid locations in east Africa. *Agricultural and Forest Meteorology* 117. 23-37.

Baudron, F., Andersson, J.A., Corbeels, M., and Giller, K. 2011. Failing to yield? Ploughs, conservation agriculture and the problem of agricultural intensification: an example from the Zambezi Valley, Zimbabwe. *Journal of Development Studies* 14(25). 1-28.

Belder, P., Rohrbach, D., Twomlow, S., and Senzanje, A. 2007. Can drip irrigation improve the livelihoods of smallholders? Lessons learned from Zimbabwe. Global theme on Agroecosystems Report Number 33. International Crops Research Institute for the Semi-Arid Tropics (ICRISAT).

Berry, S. 2002. Debating the land question in Africa. *Comparative Studies in Society and History* 44(4). 638-668.

Bird, K., and Shepherd, A. 2003. Chronic poverty in semi-arid Zimbabwe. Overseas Development Institute CPRC Working Paper Number 18.

Biswas, A.K. 2004. Integrated Water Resources Management: A Reassessment A Water Forum Contribution. W*ater International* 29(2). 248–256.

Blaikie, P. 1985. *The political economy of soil erosion in developing countries.* Longman, London.

Boelens, R. and Davila, G. (eds.). 1998. *Searching for equity: Conceptions of justice and equity in peasant irrigation.* Assen, Van Gorcum,

Bolding, A. 1999. Caught in the catchment: past, present and future of Nyanyadzi water management. In Manzungu, E., Senzanje, A., and Van der Zaag, P. (eds). *Water for agriculture in Zimbabwe: policy and management options for the smallholder sector.* University of Zimbabwe Publications, Harare. 123-152.

Bolding, A., Manzungu, E. and van der Zaag, P. 1999. A realistic approach to water reform in Zimbabwe, in Manzungu, E., Senzanje, A. and Van der Zaag, P. (eds.), *Water for agriculture in Zimbabwe: Policy and management options for the smallholder sector*, University of Zimbabwe Publications, Harare. 225-53.

Bolding, A. (2004) *In hot water: a study on sociotechnical intervention models and practices of water use in smallholder agriculture, Nyanyadzi Catchment, Zimbabwe*, PhD Thesis, University of Wageningen.

Bond, P. 2001. Challenges for the provision of social services in the 'New' South Africa: the case of residual water apartheid,' Paper presented to the Southern African Regional Institute of Policy

Studies 2001 Colloquium on Social Policy and Development in Southern Africa, Harare, 25 September 2001.

Bond, P., and Manyanya, M. 2003. *Zimbabwe's plunge: exhausted nationalism, neo-liberalism and the search for social justice*. University of Natal Press, Pietermaritzburg.

Boserup, E. 1981. *Population and technological change: a study of long term trends*. University of Chicago Press. Chicago.

Bromley, D.W. 1989. *Economic interests and institutions: the conceptual foundations of public policy*. Blackwell, Oxford.

Bulawayo Engineering Services Department; NORPLAN AS; Stewart Scott Consultants; and CNM and Partners. 2001. *Bulawayo Water Conservation and Sector Services Up-grading Project: Final Report* Volume 1, 30 April 2001.

Burawoy, M, Burton, A., Ferguson, A.A., Fox, K.J., Gamson, J., Gartrell, N., Hurst, L., Salzinger, J., and Ui, S. 1991. *Ethnography unbound: Power and resistance in the modern metropolis*. University of California Press, Berkley. Los Angeles

Burawoy, M. 1998. The extended case method. *Sociological Theory* 16. 4-33.

Butterworth, J., Warner, J., Moriarty, P., Smits, S. and Batchelor, C. 2010. Finding practical approaches to Integrated Water Resources Management. *Water Alternatives* 3(1): 68-81

Campbell, B., Sayer, J. A., Frost, P., Vermeulen, S., Ruiz Pirez, M., Cunningham, A., and Prabhu. R. 2001. Assessing the performance of natural resource systems. *Conservation Ecology* 5(2): 22. [online] URL: http://www.consecol.org/vol5/iss2/art22

Campbell, H. 2003. *Reclaiming Zimbabwe: the exhaustion of the patriarchal model of liberation.* David Philip Publishers. Claremont.

Campbell, S.D.G., and Pitfield, P.E.J. 1991. Training small-scale gold miners in Zimbabwe. *AGID News* 66, May 1991.

Cardwell, H.E., Cole, R.A., Cartwright, L.A., and Martin, L.A. 2006. Integrated water resources management: definitions and conceptual musings. *Journal of Contemporary Water Research and Education* 135. 8-18.

Catholic Commission for Justice and Peace in Zimbabwe. 1997 *Report on the 80s atrocities in Matabeleland and the Midlands.*

Central Statistical Office. 2002. *Census 2002 National Report*, Harare.

Chambers, R., and Conway, G. 1992. Sustainable rural livelihoods: practical concepts for the 21st century, *IDS Discussion Paper 296*, Institute of Development Studies, Brighton, Sussex.

Chambers, R. 2004. *Ideas for development: reflecting forwards.* IDS Working Paper 238, Institute of Development Studies, Brighton, Sussex, UK.

Chenje, M., Sola, L., and Paleczny, D. (eds). 1998. *The state of Zimbabwe's environment*, Government of the Republic of Zimbabwe, Ministry of Mines, Environment and Tourism, Harare.

Chereni, A. 2007. The problem of institutional fit in IWRM: a case of Zimbabwe's Mazowe Catchment. *Physics and Chemistry of the Earth* 32. 1246-1256.

Chidenga, E.E. 2003. *Leveraging water delivery: irrigation technology choices and operations and maintenance in smallholder systems in Zimbabwe*. PhD Thesis, University of Wageningen.

Chigwenya, A. 2010. Institutional decay and environmental degradation in Ward 29 Gutu West, Zimbabwe. *Journal of Sustainable Development in Africa* 12 (1).

Chikozho C. 2008. Stakeholder Participatory Processes and Dialogue Platforms in the Mazowe River Catchment, Zimbabwe. *African Studies Review Journal* 10. University of Florida, Washington D.C.

Chikowero, M. 2007. Subalternating currents: electrification and power politics in Bulawayo, colonial Zimbabwe, 1894-1939. *Journal of Southern African Studies* 33(2). 287-306.

Cleaver, F. 1990. Community *maintenance of handpumps: report to the National Action Committee Water and Sanitation Studies Fund*. University of Zimbabwe, Harare.

Cleaver, F. 2000. Moral ecological rationality, institutions and the management of natural resources. *Development and Change* 31(2). 361-383.

Cleaver, F. 2002. Reinventing institutions: bricolage and the social embeddedness of natural resources management. *The European Journal of Development Research* 14(2). 11-30.

Cleaver, F. 2005. _The inequality of social capital and the reproduction of chronic poverty. *World Development* 33 (6). 893-906.

Cleaver, F., and Franks, T. 2008. Distilling or diluting? Negotiating the water research-policy interface. *Water Alternatives* 1(1). 157-177.

Colvin, J., Ballim, F., Chimbuya, S., Everard, M., Goss, J., Klarenberg, G.G., Ndlovu, S., Ncala, D., and Weston, D. 2008. Building capacity for co-operative governance as a basis for integrated water resources managing in the Inkomati and Mvoti catchments, South Africa. *Water SA* 34 (6) 681-690.

Conca, K. 2006. *Governing water: contentious transnational politics and global institutions building.* The MIT Press, Cambridge, MA.

Consultative Group on International Agricultural Research (CGIAR) Task Force on Integrated Natural Resources Management. 2001. Integrated Management for sustainable agriculture, forestry and fisheries, Report of a workshop held in Cali, Colombia, 28-31 August 2001.

Conyers, D., and Hill, P. 1984. *An introduction to development planning in the Third World*, John Wiley and Sons, New York.

Cooke, B., and Kothari, U. 2001. *The tyranny of participation.* ZED Books. London.

Coward E.W. Jr. 1980. Irrigation development: institutional and organizational issues. In Coward E.W. Jr. (ed.) *Irrigation and Agricultural Development in Asia: perspectives from the social sciences.* 15-27. Cornell University Press, Ithaca, New York.

Cromby, J., and David Nightingale. 1999. What's wrong with social constructionism? In D.J.Nightingale and J.Cromby (eds.) *Social Constructionist Psychology: a critical analysis of theory and practice.* Open University Press, Buckingham. 1-20.

CSO. 2002. *Zimbabwe Population Census Report.* Central Statistical Office. Government Printers, Harare.

Davis, K., Oxley, S., and Evans, A. 2006. *Crop protection in reduced tillage systems.* SAC Technical Note 580.

De Groen, M. M., and Savenije, H. H. G. 2006. A monthly interception equation based on the statistical characteristics of daily rainfall. *Water Resources Research* 42, W12417, doi:10.1029/2006WR005013.

Dent, M. 2011. *Why do we need CMAs?* CMA Leadership Letter 118

Department of Water Affairs and Forestry (DWAF). 2002. *Strategic Plan: Multi-year 2002/3-2004/5.* Department of Water Affairs and Forestry, Republic of South Africa.

Department for Environment Food and Rural Affairs. 2011. *Weed management. Crop establishment- a management guide.* University of Hertfordshire.

Demusz, K. 1998. From Relief to Development: negotiating the continuum on theThai-Burmese1 border. *Journal of Refugee Studies II(3).*

Derman, B., Hellum, A., Manzungu, E., Sithole, P., and Machiridza, R. 2007. Intersections of law, human rights and water management: implications for rural livelihoods. In van Koppen, B., Giordano, M., and Butterworth, J. (eds) *Community-based water law and water resources management reform in developing countries*, Cambridge, Cabi International. 248-271.

Diemer, G., and Huibers, F.P. 1996. Introduction. In Diemer, G., and Huibers, F.P. (eds), *Crops, people and irrigation: water allocation practices of farmers and engineers*, London, Intermediate Technology Publications. 1-10.

Donkor, S.M.K. 1991. *A project management model based on observation-response practices for small scale irrigation scheme.* PhD Thesis, Colorado State University, Fort Collins, The United States of America.

Dougherty, T. 1997. Realities of stakeholder participation in water resources management: the case of the Mazowe Catchment Pilot Project. In Nhira, C., and Derman, B. (eds) *Towards reforming the institutional and legal basis of the water sector in Zimbabwe: current weaknesses, recent initiatives and their operational problems.* CASS Occasional Paper-NRM Series CPN86/1997.

Dubash, N.K.; Dupar, M.; Kothari, S. and Lissu, T. 2001. *A watershed in global governance? An independent assessment of the World Commission on Dams.* Washington, DC: World Resources Institute, Lokayan and Lawyer's Environmental Action Team.

Dube, E., Chiduza, C., and Muchaonyerwa, P. 2012. Conservation agriculture effects on soil organic matter on a Haplic Cambisol after four years of maize–oat and maize–grazing vetch rotations in South Africa *Soil and Tillage Research* 123. 21–28

Eldridge, C. 2002. Why was there no famine following the 1992 southern African drought? *IDS Bulletin* 33(4). 79-87.

Elwell H.A. 1981. A soil loss estimation technique for southern Africa. In Morgan, R.P.C. (ed). *Soil Conservation: Problems and Prospects.* John Wiley, Chichester. 281-292.

Falkenmark, M. 1998. Dilemma when entering 21st Century-rapid change but lack of sense of urgency. *Water Policy* 1. 421-436.

Farrington, J., Turton, C., and James, A.J. (eds). 1999. *Participatory watershed development: challenges for the Twenty-First Century*. Oxford University Press, New Dehli.

Fatch, J.J., Manzungu, E., and Mabiza, C., 2010. Problematising and conceptualizing local participation in transboundary water resources management. *Physics and Chemistry of the Earth* 35. 838-847.

Filho, C.M.C., and Gonçalves, R.S. (2010) The National Development Plan as a Political Economic Strategy in Evo Morales's Bolivia: Accomplishments and Limitations. *Latin American Perspectives* 37(4). 177-196

Forsyth, T., Leach, M., and Scoones, I. 1998. Poverty and environment: priorities for research and policy: an overview study. *Report prepared for the United Nations Development Programme and the European Commission*. Institute of Development Studies.

Garcia, L.E. 2008. Integrated water resources management: a small step for conceptualists, a giant step for practitioners. *Water Resources Development Volume* 24(1). 23-36.

Gearey, M., and Jeffrey, P. 2006. Concepts of legitimacy within the context of adaptive water management strategies. *Ecological Economics* Volume 60, pp. 129-137.

Gergen, K.J. 1985. The social construction movement in modern psychology. *American Psychologist* 40(3). 266-175.

Gibson, C., E. Ostrom, and T.-K. Ahn. 2000. The concept of scale and the human dimensions of global change: a survey. *Ecological Economics* 32. 217-239.

Giddens, A. 1983. Comments on the theory of structuration. Journal for the Theory of Social Behaviour 13(1). 75-80.

Giller, K.E., Witter, E., Corbeels, M., and Titonnell, P. 2009. Conservation agriculture and smallholder farming in Africa: the heretics' view. *Field Crop Research* 114(1).23-34.

Gleick, P.H. 1999. A human right to water. *Water Policy* 1(5). 487-503.

Global Water Partnership. 2000. Integrated Water Resources Management, *Technical Advisory Committee Background* Papers Number 4, Global Water Partnership, Stockholm.

GWP Technical Committee 2004 *Catalysing change: a handbook for developing Integrated Water Resources Management (IWRM) and water efficiency strategies*, Global Water Partnership, Stockholm.

Gottret, M.V., and White, D. 2001. Assessing the impact of integrated natural resources management: challenges and experiences. *Conservation Ecology* 5(2):17 [online] URL: http:www.consecol.org/vol5/iss2/art17.

Government of Zimbabwe. 2009. *Zimbabwe Millennium Development Goals: 2000-2007 Mid-term Progress Report*. A report of the Government of Zimbabwe to the United Nations

Grant, R.W. 2002. The ethics of incentives: historical origins and contemporary understandings. *Economics and Philosophy* 18(01). 111-139.

Grigg, N.S. 2008. Integrated water resources management: balancing views and improving practice. *Water International* 33(3). 279-292.

Gumbo, B., and Van der Zaag, P. 2002. Water losses and the political constraints to demand management: the case of the City of Mutare, Zimbabwe. *Journal of Physics and Chemistry of the Earth* 27. 805-813.

Gumbo, B., Juizo, D., and Van der Zaag, P. 2003. Information is a prerequisite for water demand management: experiences from four cities in Southern Africa, *Journal of Physics and Chemistry* of the Earth 28. 827-837.

Gumbo, B. 2004. IWRM friendly water policies: water demand management in Bulawayo, Zimbabwe. Final Report for GWP-SA IWRM Case Study Development IWRM Friendly Policies.

Haggblade, S., and Tembo, G. 2003. Early evidence on conservation farming in Zambia. Paper prepared for the International Workshop on 'Reconciling Rural Poverty and Resources Conservation: Identifying Relationships and Remedies.' Cornell University, Ithaca, New York. May 2-3, 2003.

Hagmann, J. 1996. Mechanical soil conservation with contour ridges: cure for, or cause of rill erosion?' *Land Degradation and Development* 7. 145-160.

Hagmann, J.R., Chuma, E., Murwira, K., Connoly, M., Ficarelli, P. 2002. Success factors in integrated natural resources management research and development: lessons from practice. *Conservation Ecology* 5(2):29.

168

Heynen, N., Kaika, M., and Swyngedouw, E. 2006. Urban political ecology: politicizing the production of urban natures' Heynen, N.; Kaika, M.; Swyngedouw, E. (eds) *In the nature of cities of cities: urban political ecology and the politics of urban metabolism*, Routledge. 1-20.

Herbst, J. 1990. *State politics in Zimbabwe*. University of Zimbabwe Publications. Harare.

Hilson, M. (2006) *Small-scale mining, rural subsistence and poverty in West Africa*, Practical Action Publishing, Warwickshire

Hoadley, M., Limpitlaw, D., and Weaver, A. 2002. Mining, minerals and sustainable development in southern Africa Volume 1. *The Report of the Regional MMSD Process* Volume 1, School of Mining Engineering, University of the Witwatersrand, South Africa.

Homann, S., van Rooyen, A., Moyo, T., and Nengomasha, Z. 2007. *Goat production and marketing: Baseline information for semi-arid Zimbabwe*. International Crops Research Institute for the Semi-Arid Tropics.

Hove, L. 2006. Agricultural technology transfer under relief and recovery programs in Zimbabwe: are NGOs meeting the challenge? *ICRISAT Briefing Note* 6.

Huckle, J. and Martin, A. 2001. *Environments in a Changing World*. Prentice Hall. London.

Hyden, L-C. 2008. Narratives in illness: a methodological note. *Qualitative Sociological Review* IV(3).

ICWE. 1992. *The Dublin Statement and Report of the Conference on Water and the Environment: Development Issues for the 21st C*, 26-31 January 1992, Dublin.

International Food Policy Research Institute (IFPRI). 2002. *Green revolution: curse or blessing?* Washington.

Jackson, J.C., and Collier, P.C. 1988. Incomes, poverty and food security in the communal lands of Zimbabwe. *International Institute of Social Studies Working Paper Series Number 44/RUP Occasional* Paper Number 11, Department of Rural and Urban Planning, University of Zimbabwe.

Jayne, T.S., Chisvo, M., Rukuni, M., and Masanganise, P. 2006. Zimbabwe's food insecurity paradox: hunger amid potential. In Rukuni, M., Tawonezvi, P., Eicher, C., Munyuki-Hungwe, M., and Matondi, P. (eds). *Zimbabwe's agricultural revolution revisited*. University of Zimbabwe Publications, Harare. 525-542.

Jeffrey, P., and Gearey, M. 2006. Integrated water resources management: lost on the road from ambition to realization? *Water Science and Technology* 53(1). 1-8.

Jones, S., Matiza, G., Mlauzi, B., and Wiggins, S. 2005. *Zimbabwe: Protracted Relief Programme (PRP): output to purpose review*

Jonker, L. 2007. Integrated water resources management: the theory-praxis nexus, a South African perspective. *Physics and Chemistry of the Earth 32*. 1257-1263.

Kaarsholm, P. 1995. 'Si Ye Pambili-Which way forward?: Urban development, culture and politics in Bulawayo. *Journal of Southern African Studies* 21(2). 225-245.

Kabell, T, C. 1986. Assessment of Design Flood Hydrographs. *The Zimbabwe Engineer* .24 (1).

Kassama, A., Friedrichb, T., Derpschc, R., Lahmard, R., Mrabete, R., Baschf, G., González-Sánchezg, E.J., Serraj, R. 2012 Conservation agriculture in the dry Mediterranean climate. *Field Crops Research* 132. 7–17.

Katsi, L., Siwadi, J., Guzha, E., Makoni, F.S., Smits, S. 2007. Assessment of factors which affect multiple uses of water sources at household level in rural Zimbabwe: a case study of Marondera, Murehwa and Uzumba Maramba Pfungwe districts. *Physics and Chemistry of the Earth* 32. 1157-1166.

Keen, M. 2003. Integrated water management in the South Pacific: policy, institutional and socio-cultural dimensions. *Water Policy* 5. 147-164.

Killick, T. 1983. Development planning in Africa: experiences, weaknesses and prescriptions. *Development Policy Review* 1(1). 47-76.

Koudstaal, R., Rijsberman, F.R., and Savenije, H. 1992. Water and sustainable development,' *Natural Resources Forum* November 1992.

Krammer, E. 1997. The early years: extension services in peasant agriculture in colonial Zimbabwe, 1925-1929. *Zambezia* XXIV. 159-179.

Kramarenko, V., Engstrom, L., Verdier, G., Fernandez, G., Oppers, S. E., Hughes, R., McHugh, J., and Coats, W. 2010. *Zimbabwe: challenges and policy options after hyperinflation*. International Monetary Fund. Washington DC.

Kujinga, K., and Manzungu, E. 2004. Enduring contestations: stakeholder strategic action in water resources management in the Save catchment area, Eastern Zimbabwe. *Eastern African Social Science Review* XX(1). 67-91.

169

Kujinga, K., and Jonker, L. 2006. An analysis of stakeholder knowledge about water governance transformation in Zimbabwe. *Physics and Chemistry of the Earth* 31. 690-698.

Kurtz, H. E. 2003. Scale frames and counter-scale frames: constructing the problem of environmental injustice. *Political Geography* 22. 887-916.

Lankford, B. 2009. Viewpoint – The right irrigation? Policy directions for agricultural water management in sub-Saharan Africa. *Water Alternatives* 2(3): 476-480

Lankford, B.A., Merrey, D.J., Cour, J., and Hepworth, N. 2007. *From integrated to expedient: an adaptive framework for river basin management in developing countries.* International Water Management Institute Research Report 110. Colombo.

Lankford, B. 2010. Watering Africa's sleeping giant? *Future Agricultures/SOAS Report.*

Lankford, B. and Hepworth, N. 2010. The cathedral and the bazaar: Monocentric and polycentric river basin management. *Water Alternatives* 3(1): 82-101

Latham, C.J.K. 2002. Manyame Catchment Council: a review of the reform of the water sector in Zimbabwe. *Physics and Chemistry of the Earth* 27. 907-918.

Lebel, L., P. Garden, and M. Imamura. 2005. The politics of scale, position, and place in the governance of water resources in the Mekong region. *Ecology and Society* 10(2): 18. [online] URL: http://www.ecologyandsociety.org/vol10/iss2/art18/

Long, N., and Van der Ploeg, J.D. 1989. Demythologizing planned intervention: an actor perspective. *Sociologia Ruralis* XXIX (3-4). 226-249.

Long, N. 2001. *Development sociology: actor perspectives.* Routledge, London.

Love, D., Uhlenbrook, S., Corzo-Perez, G., Twomlow, S., and Van der Zaag, P. 2010. Rainfall-interception-evaporation-runoff relationships in a semi-arid catchment, northern Limpopo basin, Zimbabwe. *Hydrological Sciences Journal* 55(5). 687-703.

Lovell, C., A. Mandondo, and P. Moriarty. 2002. The question of scale in integrated natural resource management. *Conservation Ecology* 5(2): 25.

Lutz, E., Pagiola, S., and Reiche, C. (eds). 1994. Economic and Institutional Analyses of Soil Conservation Projects in Central America and the Caribbean. *World Bank Environment* Paper No.8. Washington, DC: The World Bank.

Mabiza, C., Taru, P., Gandidzanwa, C.P., and Kuvarega, A.T. 2006. An Evaluation of Stakeholder Participation in Catchment Planning: Case study of Manyame, Catchment Area, Zimbabwe. *African Journal of Sustainable Development in Africa* 8 (1). 37-52.

Machingambi, M., and Manzungu, E. 2003. An evaluation of rural communities' water use patterns and preparedness to manage domestic water sources in Zimbabwe. *Physics and Chemistry of the Earth* 28. 1039-1046.

Madzudzo, E., and Hawkes, R. 1996. Grazing and cattle as challenges in Community Based Natural Resources Management in Bulilimamangwe District of Zimbabwe. *Zambezia* XXIII(I). 1-18.

Magadlela, D. 2000. *Irrigating lives: development intervention and dynamics of social relationships in an irrigation project.* PhD Thesis (published) Wageningen University, The Netherlands.

Maisiri, N., Senzanje, A., Rockström, J. and Twomlow, S.J. 2005. On farm evaluation of the effect of low cost drip irrigation on water and crop productivity compared to conventional surface irrigation system. *Physics and Chemistry of the Earth* 30. 783-791.

Makadho, J.M. 1994. *An analysis of water management performance in smallholder irrigation schemes.* PhD Thesis. University of Zimbabwe, Harare.

Makadho, J. 1996. Irrigation timeliness indicators and application in smallholder irrigation systems in Zimbabwe. *Irrigation Drainage Systems.* 10. 367-376.

Makochekanwa, A., and Kwaramba M. 2009. *State Fragility: Zimbabwe's horrific journey in the new millennium.* Accra: A Research Paper Presented at the European Report on Development's (ERD). 1-36.

Makoni, F.S., Manase, G., and Ndamba, J. 2004. Patterns of domestic water use in rural areas of Zimbabwe: gender roles and realities. *Journal of Physics and Chemistry of the Earth* 29. 1291-1294.

Makumbe, J. 1996. *Participatory development: the case of Zimbabwe.* Harare, University of Zimbabwe Publications.

Makurira, H., Mugumo, M., 2006. Water Sector Reforms in Zimbabwe: The Importance of policy and institutional coordination on implementation. Proceedings of the African Regional workshop on watershed management, Nairobi, October 2005. FAO (2006). 167-174.

Makwarimba, E., and Vincent, L.F. 2004. Job satisfaction and the organisational lifeworlds of extension workers in irrigation in Manicaland. In Moll, H.A.J., Leeuwis, C., Manzungu, E., and Vincent, L.F. (eds) *Agrarian institutions between policies and local action: experiences from Zimbabwe*, Weaver Press, Harare. 241-280.

Mandondo, A. 2000. Situating Zimbabwe's natural resource governance in history. *CIFOR Occasional Paper 32.*

Manyanga, M. 2006. *Resilient landscapes: Socio-environmental dynamics in the Shashi-Limpopo Basin, Southern Zimbabwe c. Ad 800 to the present.* PhD thesis. Uppsala University, Sweden.

Manzungu, E., and P. van der Zaag (eds). 1996. *The practice of smallholder irrigation: case studies from Zimbabwe.* University of Zimbabwe Press, Harare.

Manzungu, E. 1999. *Strategies of smallholder irrigation management in Zimbabwe.* PhD Thesis, University of Wageningen. The Netherlands

Manzungu, E. 2001. A lost opportunity: an analysis of the water reform debate in the fourth parliament of Zimbabwe. *Zambezia*, 28 (1). 97-120.

Manzungu, E. 2002. More than a headcount: towards strategic stakeholder representation in catchment management in South Africa and Zimbabwe. *Physics and Chemistry of the Earth* 27, pp. 927-934.

Manzungu, E. 2003. *Towards sustainable water resources and irrigation development in the post Fast Track Land Reform Land Reform era in Zimbabwe.* In Report of The Presidential Land Review Committee Volume 2 Special Studies. 56-91.

Manzungu, E. 2004. Public institutions in smallholder irrigation in Zimbabwe. In Moll, H.A.J., Leeuwis, C., Manzungu, E., and Vincent, L.F. (eds). *Agrarian institutions between policies and local action: experiences from Zimbabwe.* Weaver Press. Harare. 27-56.

Manzungu, E. 2004. Water for all: improving water resources governance in Southern Africa. *Gatekeeper Series* 113.

Manzungu, E., and Machiridza, R. (2005) Economic-legal ideology and water management in Zimbabwe: implications for smallholder agriculture. Paper presented at the International Workshop on' African water laws: plural legislative frameworks for rural water management in Africa, Johannesburg, South Africa, 26-28 January 2005.

Manzungu, E. and R. Machiridza. 2009. Economic-legal ideology and water management in Zimbabwe. *Economics, Management and Financial Markets* 4(1).

Manzungu, E. Sithole, P., Tapela, B. and van Koppen, 2009. Phases and interfaces: national and local investments in Sekororo communal lands, South Africa. *Economics, Management and Financial Markets* 4(4).

Manzungu, E., Mpho, T., and Mdange-Mpale, A. 2009. Continuing discontinuities: local and state perspectives on cattle and water resource management in Botswana. *Water Alternatives* 2(2). 205-224.

Mapedza, E. 2007. Keeping CAMPFIRE going: political uncertainty and natural resources management in Zimbabwe. *Gatekeeper Series* 133.

Maphosa, B. 1994. Lessons from the 1992 Drought in Zimbabwe: The Quest for Alternative Food Policies. *Nordic Journal of African Studies* 3(1). 53–58.

Maphosa, B. 1994. Lessons from the 1992 drought in Zimbabwe: the quest for alternative food policies. *Nordic Journal of African Studies* 3(1). 53-58.

Maponga, O., and Ngorima, C.F. 2003. Overcoming environmental problems in the gold panning sector through legislation and education: the Zimbabwean experience. *Journal of Cleaner Production* 11. 147-157.

Mashingaidze, T.M. 2006. The Zimbabwean entrapment: an analysis of the nexus between domestic and foreign policies in a 'collapsing' militant state, 1990s-2006. *Alternatives* 5(4). 57-76.

Massa, S., and Testa, S. 2008. Innovation and SMEs: misaligned perspectives and goals among entrepreneurs, academics and policy makers. *Technovation* 28. 393-407.

Matabeleland Chamber of Industries. 1994. '100 years of industry in Bulawayo.' Unpublished brochure/pamphlet

Matondi, P., and Munyuki-Hungwe, M.N. 2006. The evolution of agricultural policy: 1990-2004. In Rukuni, M., Tawonezvi, P., Eicher, C., Munyuki-Hungwe, M., and Matondi, P. (eds). *Zimbabwe's agricultural revolution revisited.* University of Zimbabwe Publications, Harare. 63-97.

Matsika, N. 1996. Challenges of independence: managing technical and social worlds in a farmer managed irrigation scheme. In Manzungu, E., and Van der Zaag, P. (eds). 1996. *The practice of smallholder irrigation: case studies from Zimbabwe*. University of Zimbabwe Press, Harare. 29-46.

McCaffrey, S.C., and Neville, K.J. 2009. Small capacity and big responsibilities: financial and legal implications of a human right to water for developing countries. *Georgetown International Environmental Law Review* 21 (4). 679-704.

McFarlane, M.J., and Whitlow, R. 1990. Key factors affecting the initiation and progress of gullying in dambos in parts of Zimbabwe and Malawi. *Land Degradation and Rehabilitation* 2. 215-235.

McGregor, J. 1995. Conservation, control and ecological change: the politics and ecology of colonial conservation in Shurugwi, Zimbabwe. *Environment and History* 1(3). 253-279.

Medema, W., and Jeffrey, P. 2005. IWRM and adaptive management: synergy or conflict?' *NeWater Report* Series Number 7.

Mehta, L. 2001. The Manufacture of Popular Perceptions of Scarcity: Dams and water related narratives in Gujarat, India. *World Development 29(12)*. 2025-2041

Meinzen-Dick, R., and Nkonya, L. 2007. Understanding legal pluralism in water and land rights: lessons from Africa and Asia. In van Koppen, B., Giordano, M., and Butterworth, J. (eds), *Community-based water law and water resources management reform in developing countries,* Cambridge, Cabi International. 12-27.

Merrey, D. J., Drechsel, P., Penning de Vries, P., and H. Sally. 2005. Integrating 'Livelihoods' into Integrated Water Resources Management: Taking the Integration Paradigm to its Logical Next Step for Developing Countries. *Regional and Environmental Change* 5:197-204.

Merrey, D. J. 2007 Is normative Integrated Water Resources Management implementable? *bscw.ihe.nl/pub/nj_bscw.cgi/S4b4f24b6/d2606885/Merrey.pdf (26/07/2011)*

Merrey, D.J. 2008. Is normative integrated water resources management implementable? Charting a course with lessons from Southern Africa. *Journal of Physics and Chemistry of the Earth* 33. 899-905.

Merrey, D.J. 2009. African models for transnational river basin organisations in Africa: an unexplored dimension. *Water Alternatives* 2 (2). 183-204.

Merrey, D.J. 2009. African models for transnational river basin organisations in Africa: an unexplored dimension. *Water Alternatives* 2(2). 183-204.

Milward, H.B., and Provan, K.G. 2000. Governing the hollow state. *Journal of Public Administration Research and Theory* 2. 359-379.

Minerals, Mining and Sustainable Development (MMSD). 2002. *Breaking new ground: minerals, mining and sustainable development*, Earthscan, London.

Mitchell, B. 1990. *Integrated water management: international experiences and perspectives*, Belhaven Press, London.

Mitchell, B. 2005. Integrated water resources management; institutional arrangements, and land-use planning,' *Environment and Planning A* 37. 1335-1352.

Mitchell, B. 2006. IWRM in practice: lessons from Canadian experiences. *Journal of Contemporary Water Research and Education* 135. 51-55.

Mkandla, N., Van der Zaag, P., and Sibanda, P. 2005. Bulawayo water supplies: Sustainable alternatives for the next decade. *Physics and Chemistry of the Earth* 30. 935–942.

Moench, M. Caspari, E., and Dixit, A. (eds.). 1999. *Rethinking the mosaic: Investigations into local water management. Kathmandu: Nepal Water Conservation Foundation and Boulder*. Institute for Social and Environmental Transition.

Molle, F. 2008. Nirvana concepts, narratives and policy models: Insight from the water sector. *Water Alternatives* 1(1). 131-156.

Molle, F. 2009. River basin planning and management: the social life of a concept. *Geoforum* 40. 484-494.

Mollinga, P.P. (ed). 2000. *Water for food and rural development. Approaches and initiatives in South Asia*. Sage, New Delhi.

Mollinga, P.P.; Meinzen-Dick, R.S.; and Merrey, D.J. 2007. Politics, plurality and problemsheds: a strategic approach for reform of agricultural water resources management. *Development Policy Review* 25(6). 699-719.

Mollinga, P.P. 2008. Water, politics and development: Framing a political sociology of water resources management. *Water Alternatives* 1(1): 7-23

Moore, D.; Dore, J. and Gyawali, D. 2010. The World Commission on Dams + 10: Revisiting the large dam controversy. *Water Alternatives* 3(2): 3-13

Moriarty, P., Butterworth, J., Van Koppen, B., and Soussan, J. 2004. Water, poverty and productive uses of water at the household level. In Moriarty, P., Butterworth, J., Van Koppen, B. (eds). *Beyond domestic: case studies on poverty and productive uses of water at the household level.* Technical Paper Series 41. IRC International Water and Sanitation Centre, Delft. 19-48.

Moriarty, P., Batchelor, C., Laban, P., and Fahmu, H. 2010. Developing a practical approach to 'Light IWRM' in the Middle East. *Water Alternatives* 3(1). 122-136.

Moyce, W., Mangeya, P., Owen, R. and Love, D. 2006. Alluvial aquifers in the Mzingwane Catchment: their distribution, properties, current usage and potential expansion. *Physics and Chemistry of the Earth* 31. 988-994.

Moyo, M. 2004. Participation dynamics in integrated water management in the Mazowe watershed. In Moll, H.A.J., Leeuwis, C., Manzungu, E., and Vincent, L.F. (eds) *Agrarian institutions between policies and local action: experiences from Zimbabwe.* Weaver Press, Harare. 109-126.

Moyo, S. 2001. Building landscapes: village development in Zimbabwe. *Land Degradation and Development* 12. 217-224.

MRRWD (Ministry of Rural Resources and Water Development), n.d. *Towards integrated water resources management: Water resources management strategy in Zimbabwe.* Harare, Zimbabwe.

Mtsi, S., and Nicol, A. 2003. Caught in the act: New stakeholders, decentralisation and water management process in Zimbabwe. *Sustainable livelihoods in Southern Africa.* Institute of Development Studies Research Paper 14.

Mtsi, S. 2008. *Zimbabwe's Water Reform and Effects on Local Level Water Management Processes and Rural Livelihoods: Evidence from Lower Save East Sub-Catchment,* PhD Thesis (unpublished), University of Manchester.

Mtsi, S. 2011. Water reforms during the crisis and beyond Understanding policy and political challenges of reforming the water sector in Zimbabwe. Working Paper 333. Overseas Development Institute (ODI).

Mugabe, F.T., Chivizhe, J., Hungwe, C. 2008. *Quantitative assessment of the effectiveness of drip irrigation kits in alleviating food shortages and its success in Zimbabwe: a case study of Gweru and Bikita districts.* Final Report. FANRPAN.

Mulwafu, W.O., and Msosa, H.K. 2005. IWRM and poverty reduction in Malawi: a socio-economic analysis. *Physics and Chemistry of the Earth* 30. 961-967.

Muller, A.M. 2006. Sustaining the right to water in South Africa. *Human Development Report 2006.* Human Development Report Office, Occasional Paper.

Munamati, M. and Nyagumbo, I. 2010. *In situ* rainwater harvesting using dead level contours in semi-arid southern Zimbabwe: insights on the role of socio-economic factors on performance and effectiveness in Gwanda District. *Physics and Chemistry of the Earth* 35. 699-705.

Mupangwa, W., Twomlow, S., and Walker, S. 2008. The influence of conservation tillage methods on soil water regimes in semi-arid southern Zimbabwe. *Physics and Chemistry of the Earth* 33. 762-767.

Musemwa, M. 2008. Early struggles over water: from private to public water in the city of Bulawayo, Zimbabwe, 1894-1924. *Journal of Southern African Studies* 34(4).

Mutenheri, F. 2009. Rational or disjointed: grappling with philosophical contradictions in 21st Century Economic Policy Planning and Development in Zimbabwe. *Journal of Sustainable Development in Africa* 10(4).

Mutiro, J., Makurira, H., Senzanje, A., and Mul, M.L. 2006. Water productivity analysis for smallholder rainfed systems: a case study of Makanya Catchment, Tanzania. *Physics and Chemistry of the Earth* 31. 901-909.

Myers, N. 2005. Environmental refugees: an emergent security issue. Paper presented at the 13th Economic Forum. Prague, 23-27 May 2005.

Nemarundwe, N. 2003. *Negotiating resource access: Institutional arrangements for woodlands and water use in southern Zimbabwe.* PhD Thesis, Swedish University of Agricultural Studies, Uppsala, Sweden.

Newson, M. 1997. *Land, water and development: sustainable management of river basin systems* (2nd Edition). Routledge, London.

Ngigi, S.N., Rockström, J., and Savenije, H.H.G. 2008. Assessment of rainwater retention in agricultural land and crop yield increase due to conservation tillage in Ewaso Ng'iro River Basin, Kenya. *Physics and Chemistry of the Earth* 31. 910-918.

Nkala, P., Mango, N., Corbeels, M., Jan Veldwisch, G., and Huising, J. 2011 The conundrum of conservation agriculture and livelihoods in Southern Africa. In *African Journal of Agricultural Research* Vol. 6(24). 5520-5528.

Nkonyo, E., Pender, J., Kaizzi, K.C., Kato, E., Magarura, S., Ssali, H., and Muwonge, J. 2008. *Linkages between land management, land degradation and poverty in Sub-Saharan Africa: the case of Uganda*. International Food Policy Research Institute Research Report 159.

NORPLAN A.S in association with Stewart Scott Consultants and CNM & Partners (2001) Bulawayo Water Conservation and Sector Services Up-grading Project, Final Report Volume 1.

North, D. 1990. *Institutions, institutional changes and economic performance*. Cambridge University Press, Cambridge, MA.

Nyamudeza, P. 1999. Agronomic practices for the low rainfall regions of Zimbabwe. In Manzungu, E., Senzanje, A., and van der Zaag, P. (eds.). *Water for agriculture in Zimbabwe: policy and management options for the smallholder sector*. University of Zimbabwe Publications, Harare. 49-63.

Oldrieve, B. 1993. *Conservation farming for communal, small-scale, resettlement and cooperative farmers of Zimbabwe: a farm management handbook*. Harare: Rio Tinto Foundation.

Olleta, A. 2005. The World Bank's influence on water privatisation in Argentina: the experience of the City of Buenos Aires. *International Environmental Law Research Centre (IELRC)* Working Paper 2007-02.

Ostrom, E. 1990. *Governing the commons: the evolution of institutions for collective action*. Cambridge University Press, Cambridge

Ostrom, E., Schroeder, L., Wynne, S.G. 1993. *Institutional incentives and sustainable development: Infrastructure policies in perspective*, Westview Press, Boulder.

Ove Arup and Partners. 2001. City of Bulawayo: study on the regulatory framework for water supply and sanitation in Bulawayo, Final Report.

Owen, I.R. 1995. El construccionismo social y la teoria, practica e investigacion en psicoterapia: Un manifiesto psicologia fenomenologica. [Social constructionism and the theory, practice and research of psychotherapy: A phenomenological psychology manifesto.] (Trans I. Caro). *Boletin de Psicologia* 46. 161-186.

Owen, R., Verbeek, K., Jackson, J., and Steenhuis, T. (eds). 1995. *Dambo farming in Zimbabwe: water management, cropping and soil potentials for smallholder farming in the wetlands*. University of Zimbabwe Publications, Harare.

Page, S.L.J., and Page, H.E. 1991. Western hegemony over African agriculture in Southern Rhodesia and its continuing threat to food security in independent Zimbabwe. *Agriculture and Human Values* 8 (4) pp. 3-18.

Parfitt, T. 2007. The ambiguity of participation: a qualified defence of participatory development. *Third World Quarterly* 25(3). 537-555.

Pazvakavambwa, S. 1994. Agricultural extension. In Rukuni, M., and Eicher, C.K. (eds) *Zimbabwe's agricultural revolution*. University of Zimbabwe Publications, Harare. 104-113.

Petit, O., and Baron, C. 2009. Integrated water resources management: from general principles to its implementation by the state: the case of Burkina Faso. *Natural Resources Forum* 33. 49-59.

Pirages, D. 2011. Ecological security: a framework for analysing non-traditional security issues. In Pirages, D., Sobhan, F., VanDeveer, S.D., and Li, L. (eds). *Ecological and non-traditional security challenges in South Asia*. The National Bureau of Asian Research. NBR Special Report Number 28. Washington. 1-12.

Pretty, J.N., and Shah, P. 1997. Making soil and water conservation sustainable: from coercion and control to partnerships and participation. *Land Degradation and Development* 8. 39-58

Qaddumi, H. 2008. Practical approaches to transboundary water benefit sharing. *Overseas Development Institute (ODI)* Working Paper 292.

Ramirez-Sanchez, S. 2006. Finding emotions in the drama of the commons: a multi-relational and multi-level analysis of the access to fishery resources in the Loreto Bay Marine Park, Baja, California Sur, Mexico,' Paper presented at The survival of the commons: mounting challenges and new realities the 12th Biannual Conference of the Association for the Study of Common Property (IASCP), Ubud, Bali, Indonesia, 19-23 June 2006.

Ranger, T.O. 1985. *Peasant consciousness and guerrilla war in Zimbabwe: a comparative study*. James Currey Ltd. London.

Ranger, T.O. 1999. *Voices from the rocks: nature, culture and history in the Matopos Hills of Zimbabwe*. Baobab Books, Harare.

Reckwitz, A. 2002. Towards a theory of social practices: a development in culturalist theorizing. *European Journal of Social Theory* 5 (2). 243-263.

Ribot, J.C., and Peluso, N. 2003. A theory of access. *Rural Sociology* 68(2). 153-181.

Richardson, C.J. 2007. How much did droughts matter? Linking rainfall and GDP growth in Zimbabwe. *African Affairs* 106(424). 463–478.

Rijsberman, F. 2003. Can development of water resources reduce poverty? *Water Policy* 5. 399-412.

Rimmon-Kenan, S. 2006. Concepts of narrative. In Hyvarinen, M., Korhonen, A., and Mykkanen, J. (eds). *The travelling concept of narrative*: *Studies Across Disciplines in the Humanities and Social Sciences 1*. Helsinki Collegium for Advanced Studies. Helsinki.

Robinson, P., Mathew, B., and Proudfoot, D. 2004. Productive water strategies for poverty reduction in Zimbabwe. In Moriarty, P., Butterworth, J., and Van Koppen, B. (eds). *Beyond domestic: case studies on poverty and productive uses of water at the household level*. Technical Paper Series 41. IRC International Water and Sanitation Center. Delft. 173-198.

Rockström, J., Jansson, P.E., Barron, J. 1998. Seasonal rainfall partitioning under runon and runoff conditions on sandy soil in Niger: on farm measurements and water balance modelling. *Journal of Hydrology* 210. 68-92.

Rockström, J. 2000. Water resources management in smallholder farms in Eastern and Southern Africa: an overview. *Physics and Chemistry of the Earth (B)* 25(3). 275-283.

Rockström, J., Barron, J., and Fox, P. 2002. Rainwater management for increased productivity among smallholder farmers in drought prone environments. *Physics and Chemistry of the Earth* 27. 949-959.

Rukuni, M. 1988.The evolution of smallholder irrigation policy in Zimbabwe: 1928-1986. *Irrigation Drainage Systems* 2. 199-210.

Rukuni, M. 1994. *Report of the Commission of Inquiry into Appropriate Agricultural Land Tenure Systems*. Volume II: Technical Reports. Government Printers, Harare.

Rukuni, M., and Jayne, T.S. 1995. Alleviating hunger in Zimbabwe: towards a national food security strategy. Supplement to *Zambezia*.

Rukuni, M. 2006. Revisiting Zimbabwe's agricultural revolution. In Rukuni, M., Tawonezvi, P., Eicher, C., Munyuki-Hungwe, M., and Matondi, P. (eds). *Zimbabwe's agricultural revolution revisited*. University of Zimbabwe Publications, Harare. 1-21.

SADC *Regional Strategic Action Plan on Integrated Water Resources Development and Management: annotated strategic action plan-2005-2010*. 2005. UNDP. SADC. EU.

SADC. 2005. *Implementing the SADC Regional Strategic Action Plan for Integrated Water Resources Management (1999-2 004): lessons and best practice*. SADC, GTZ, Inwent, Gaborone.

Saravanan, V.S., McDonald, G.T., and Mollinga, P.P. 2009. Critical review of Integrated Water Resources Manangement: moving beyond polarised discourse. *Natural Resources Forum 33*. 76-86.

Savenije, H., and Van der Zaag. 2002. Water as an economic good and demand management: paradigms with pitfalls. *Water International* 27 Number 1. 98-104.

Savenije, H.H.G., and Van der Zaag, P. 2008. Integrated water resources management: Concepts and issues. *Physics and Chemistry of the Earth* 33. 290–297

Schmid, A. 1998. *Thesaurus and glossary of early warning and conflict prevention terms* (abridged version), London.

Schouten, M. 2009. *Strategy and performance of water supply and sanitation providers: Effects of two decades of neo-liberalism*, PhD Thesis (published), Erasmus Universiteit, Rotterdam.

Scott, J.C. 1985. *Weapons of the weak: everyday forms of peasant resistance*. Yale University Press, New Haven.

Scott, J.C. 1998. *Seeing like a state: how certain schemes to improve the human condition have failed*. Yale University Press, New Haven.

Shah, T., Makin, I., and Sakthivadivel, R. 2005. Limits to leapfrogging: issues in transposing successful river basin management Institutions in the developed world. In Svendsen, M. (ed). *Irrigation and river basin management*. Cabi Publishing. Wallingford. 125-144.

Shah, T. and Van Koppen, B. 2006. Is India ready for IWRM? Fitting water policy to national development context' *Economic and Political Weekly*, 5 August 2006.3413-3421

Shah, T. 2007. Issues in reforming informal water economies of low income countries: examples from India and elsewhere. In Van Koppen, B., Giordano, M., Butterworth, J. (eds). *Community-based water law and water resources management reform in developing countries.* Cabi International. 65-95.

Shoko, D.S.M. 2002. Small-scale mining and alluvial gold panning within the Zambezi Basin. In Chikowore, G., Manzungu, E., Mushayavanhu, D. and Shoko, D. 2002 *Managing Common Property in an Age of Globalisation: Zimbabwean Experiences,* Weaver Press, Harare.

Sijali, I.V. 2001. *Drip irrigation: options for smallholder farmers in eastern and southern Africa.* Sida Regional Land Management Unit (RELMA), Nairobi.

Sithole, B. Fall 2001. Participation and Stakeholder Dynamics in the Water Reform Process in Zimbabwe: The Case of the Mazoe Pilot Catchment Board. *African Studies Quarterly* 3 5(3).

Smith, K.W., and Stalans, L.J. 1991. Encouraging tax compliance with positive incentives: a conceptual framework and research directions. *Law and Policy* 13(1). 35-53.

Smith, L., and Ruiters, G. 2006. The public/private conundrum of urban water: a view from South Africa. In Heynen, N., Kaika, M., Swyngedouw, E. (eds) *In the nature of cities of cities: urban political ecology and the politics of urban metabolism.* Routledge, London. 191-207.

Sneddon, C., Harris, L., Dimitrov and Ozesmi, R.S. 2002. Contested waters: conflict, scale and sustainability in aquatic socio-ecological systems. *Society and Natural Resources* 15 (8). 663-675.

Sneddon, C., and Fox, C. 2008. River basin politics and the rise of ecological and transnational democracy in Southeast Asia and Southern Africa. *Water Alternatives* 1(1). 66-88.

Snellen, W.B., and Schrevel, A. 2004. IWRM, for sustainable use of water: 50 years of international experience with the concept of integrated water resources management. Background document to the FAO/Netherlands Conference on Water for Food and Ecosystems. *Alterra Report 1143.*

Spiegel, S.J. 2009. Resource Policies and Small-Scale Gold Mining in Zimbabwe. *Resources Policy* 34: 39-44.

Sukume, C., Moyo, S., and Matondi, P.B. 2003. Farm sizes, land use and viability considerations. In *Report of The Presidential Land Review Committee* Volume 2 Special Studies. 38-55.

Svendsen, M., Wester, P., and Molle, F. 2005. Managing river basins: an institutional perspective. In Svendsen, M. (ed). *Irrigation and river basin management: options for governance and institutions.* International Water Management Institute/Cabi Publishing. Colombo. 1-18.

Swatuk, L.A. 2002. The new water architecture in southern Africa: reflections on current trends in the light of Rio + 10,' *International Affairs* 78(3).507-530.

Swatuk, L.A. 2005. Political challenges to implementing IWRM in Southern Africa. *Physics and Chemistry of the Earth* 30. 872-880

Swatuk, L.A. 2005. Political challenges to implementing IWRM in Southern Africa. *Physics and Chemistry of the Earth* 30. 872-880

Swatuk, L.A. 2008. A political economy of water in southern Africa. *Water Alternatives* 1(1). 24-47.

Swatuk, L.A. 2010. The State and Water Resources Development through the Lens of History: A South African Case Study. *Water Alternatives* 3(3). 521-536.

Swyngedouw E. 1999. Modernity and Hibridity: Nature, *Regeneracionismo,* and the Production of the Spanish Waterscape, 1890-1930. *Annals of the Association of American Geographers* 89(3). 443-465.

Swyngedouw, E. 2002. Governance, water and globalisation: a political-ecological perspective. Paper presented at the First International Conference: Meaningful interdisciplinary challenges and Opportunities for water research, School of Geography and Environment, University of Oxford, April 24-25, 2002.

Swyngedouw, E. 2006. Power, water and money: exploring the nexus. *Human Development Report 2006.* Human Development Report Office Occasional Paper, UNDP.

Tawonezvi, P.H., and Hikwa, D. 2006. Agricultural research policy,' Rukuni, M.; Tawonezvi, P., Eicher, C., Munyuki-Hungwe, M., and Matondi, P. (eds) *Zimbabwe's agricultural revolution revisited,* University of Zimbabwe Publications, Harare, pp197-214.

The Dublin Statement on Water and Sustainable Development. 1992.

Theesfeld, I. 2011. Perceived power resources in situations of collective action. *Water Alternatives* 4(1). 86-103.

Tiffen, M., Mortimore, M., and Gichuki, F. 1994. *More people, less erosion: environmental recovery in Kenya,* John Wiley and Sons, London.

Tunhuma, N.M., Keldermann, P., Love, D., Uhlenbrook S. 2007. Environmental Impact Assessment of Small Scale Resource Exploitation: the case of gold panning in Zhulube Catchment, Limpopo Basin, Zimbabwe. Conference proceedings of the 8th WATERNET / WARFSA / GWP-SA Symposium, Lusaka, Zambia, 31 Oct – 2 Nov 2007.

Turton, A. 2005. Water as a source of conflict or cooperation: the case of South Africa and its transboundary rivers. CSIR Report No: ENV-P-CONF 2005-002

Towmlow, S. and Hove, L. 2006. Is conservation farming an option for vulnerable house- holds? ICRISAT: Bulawayo.

Twomlow, S.; Love, D., and Walker, S. 2008. The nexus between integrated natural resources management and integrated water resources management in southern Africa. *Physics and Chemistry of the Earth* *33*. 889-898.

United Nations Economic Commission for Africa. 2002. Economic impact of environmental degradation in Southern Africa. United Nations Economic Commission for Africa-Southern Africa Office (ECA-SA).

UN Millennium Project. 2005. *Investing in development: A practical way to achieve the Millennium Development Goals*, United Nations Organisation, New York.

United Nations Development Programme (UNDP). 2006. *Summary Human Development Report 2006: beyond scarcity: power, poverty and the global water crisis.*

United Nations Development Programme (UNDP), (2008) *Comprehensive economic recovery in Zimbabwe: a discussion document.* UNDP Zimbabwe.

United Nations Environmental Programme (UNEP). 2007. Dams and development: relevant practices for improved decision making. UNEP-DDP Secretariat, Nairobi.

United Nations Development Programme (UNDP). 2010. *Summary: Human development report 2010.* UNDP

USAID. 2007. What is integrated water resources management? www.usaid.gov/ourwork/environment/water/what is iwrm.html (accessed 16/09/2009)

Van de Ven, A.H. 1986. Central problems in the management of innovation. *Management Science* 32(5). 590-607.

Van der Zaag, P. 1992. *Chicanery at the canal: changing practice in irrigation management in Western Mexico*, PhD Thesis, University of Wageningen

Van der Zaag, P. 2003. The bench terrace between invention and intervention: physical and political aspects of conservation technology. In Bolding, A., Mutimba, J. and Van der Zaag, P. (eds). *Agricultural intervention in Zimbabwe; new perspectives on extension,* University of Zimbabwe Publications, Harare.184-205.

Van der Zaag, P. 2005. Integrated Water Resources Management: relevant concept or irrelevant buzzword? A capacity building agenda research agenda for Southern Africa. *Physics and Chemistry of the Earth* 30. 867-871.

Van der Zaag, P. 2010. *Viewpoint –* Water variability, soil nutrient heterogeneity and market volatility – Why sub-Saharan Africa's Green Revolution will be location-specific and knowledge-intensive. *Water Alternatives* 3(1). 154-160.

Van Koppen, B., Moriarty, P., and Boelee, E. 2006. Multiple-use water services to advance the Millennium Development Goals, Research Report 98, International Water Management Institute, Colombo.

Van Koppen, B., Giordano, M., and Butterworth, J. eds. 2007. *Community based water law and water resources management reforms in developing countries.* CABI.

Vijfhuizen, C. 1998. *The people you live with. Gender identities and social practices, beliefs and power in the livelihoods of Ndau women and men in a village with an irrigation scheme in Zimbabwe.* PhD Thesis, Wageningen Universiteit, The Netherlands .

Vincent, L.F., and Manzungu, E. 2004. Water rights and water availability in the Lower Odzi watershed of the Save Catchment. In Moll, H.A.J., Leeuwis, C., Manzungu, E., and Vincent, L.F. (eds) *Agrarian institutions between policies and local action: experiences from Zimbabwe.* Weaver Press. Harare. 127-164.

Vishnudas, S., Savenije, H.H.G., Van der Zaag, P., Kumar, C.E.A., and Anil, K.R. 2008. Sustainability analysis of two participatory watersheds in Kerala. *Physics and Chemistry of the Earth* 33. 1-12.

Vishnudas, S., Savenije, H.H.G., and Van der Zaag, P. 2012. Watershed development practices for ecorestoration in a tribal area-A case study in Attappady Hills, South India. *Physics and Chemistry of the Earth* 47-48. 58-63

Warner, J. 2005. Multi-stakeholder platforms: integrating society in water resources management? *Ambiente and Sociedad* VIII (2).

Warner, J., Wester, P., and Bolding, A. 2008. Going with the flow: river basins as natural units for water management? *Water Policy* 10 Supplement 2. 121-138.

Warner, J., Lulofs, K. and Bressers, H. 2010. The fine art of boundary spanning: Making space for water in the East Netherlands. *Water Alternatives* 3(1): 137-153

Wegerich, K. 2009. Shifting to hydrological boundaries-the politics of implementation in the lower Amu Darya basin. *Physics and Chemistry of the Earth* 34. 279-288.

Wester, P., and Warner, J. 2002. River basin management reconsidered. Turton, A., and Henwood, R. (eds) *Hydropolitics in the developing world: a Southern African perspective.* African Water Issues Research Unit, Pretoria. 61-71.

Wester, P., Merrey, D.J., and De Lange, M. 2003. Boundaries of consent: stakeholder representation in river basin management in Mexico and South Africa. *World Development* 31(5). 797-812.

Whitlow, R. 1985. Conflicts in land use in Zimbabwe: political, economic and environmental perspectives. *Land Use Policy* 2(4). 309-322.

Whitlow, R. 1988. Soil erosion and conservation policy in Zimbabwe: past, present and future. *Land Use Policy* 5 (4). 419-431.

Wilson, K.B. 1995. Water used to be scattered in the landscape: local understandings of soil erosion and land use planning in southern Zimbabwe. *Environment and History* 1(3). 281-296.

Wolf, A.T. 2000. Indigenous approaches to water conflict negotiations and implications for international waters. *International Negotiation: A Journal of Theory and Practice* 5(2).

World Bank. 1993. *Water resources management: A World Bank Policy Paper.* Washington DC.

World Bank. 2004. *Water resources sector strategy: strategic directions for World* Bank engagement. Washington DC.

World Commission on Environment and Development (WCED). 1987. *Our common future: the report of the World Commission on Environment and Development.* Oxford University Press. Oxford.

World Health Organisation (WHO). 2011. *Cholera in Zimbabwe.* Global Alert and Response.

Yin, R.K. 2009. Case Study Research: Design and Methods. Fourth Edition. SAGE Publications. California

Zawe, C. 2000. *Operation and maintenance of sprinkler irrigation schemes by farmers.* MSc Thesis, Wageningen Agricultural University. The Netherlands.

Zawe, C., 2006. Reforms in Turbulent Times - A study on the theory and practice of three irrigation management policy reform models in Mashonaland, Zimbabwe. PhD Thesis, Wageningen University, The Netherlands

Zeitoun, M. and Mirumachi, N. 2008. Transboundary water interaction: reconsidering conflict and cooperation. *International Environmental Agreements* 8. 297-316.

Zikhali., P. 2010. Fast Track Land Reform Programme, tenure security and investments in soil conservation: Micro-evidence from Mazowe District in Zimbabwe Natural Resources Forum 34(2). 124–139.

Zimbabwe Vulnerability Assessment Committee (ZIMVAC). 2009. ZIMVAC rural livelihoods survey Report Number 11.

Zimbabwe. 1998a. *Water Act Chapter 20:24.* Harare.

Zimbabwe. 1998b. *Zimbabwe National Water Authority Act: Chapter 20: 25.* Harare.

Zimbabwe National Statistical Agency (ZIMSTAT). 2009. Multiple Indicator Monitoring Survey (MIMS) Preliminary Report November 2009.

Zimbabwe National Statistical Agency (ZIMSTAT). 2012. Preliminary National Census Report 2012.

Zimbabwe Parliamentary Debates. 2007a. The Senate 17(7). Thursday 13 September 2007.

Zimbabwe Parliamentary Debates. 2007b. House of Assembly 34(6). Wednesday 5th September 2007

Zinyowera, M. C., and Unganai, L. S. 1993. Drought in southern Africa: an update on the 1991 - 92 drought. *Drought Network News Int.* 4(3). 3-4. (Nebraska: International Drought Information Center, University of Nebraska).

Zwane, N., Love, D., Hoko, Z., and Shoko, D. 2006. Managing the impact of gold panning activities within the context of Integrated Water Resources Management planning in the Lower Manyame subcatchment, Zambezi Basin. *Physics and Chemistry of the Earth* 31. 848-856.

178

Appendix 1: Mzingwane Catchment questionnaire

MZINGWANE CATCHMENT

QUESTIONNAIRE FOR INSTITUTIONS, FARMS, PARKS ETC.

PREAMBLE

The management and development of water resources is now in the hands of the local communities (stakeholders) who are affected or who benefit from the resource. This is done through the Catchment Council which is a body comprising representatives of stakeholder such as Rural District Councils, miners, commercial farmers, traditional leaders and communal farmers.

The Water Act (Chapter 20:24) of 1998 requires the Catchment Council to prepare an outline water development plan for every river system in its catchment. This questionnaire is an attempt by the Mzingwane Catchment Council to gather information and the aspirations of stakeholders which will be used in preparing the plan.

Please provide the information required below to the best of your ability.

Name of institution/farm/park/property............................... Owner
District.............................Sub-Catchment.......................... Ward

How many of each type of water source are there in the farm/park/property?

____Dams___ ____Wells_____Boreholes___ _ _____Mine shafts
Well point (sand abstraction)_____Other-specify_____

Give details in the tables below:-

Table 1: Dams

Name of dam				
River				
Permit No.				
Date Constr				
Dam owner				
Grid Ref.				
Depth				
Spillway size				
Capacity (m^3)				
Silt (high/low)				

Filling				
Drying				
Major use				
Ha or m^2 (if under irr.)				
Ha or m^3 (of WS)				
Water Colour				
Taste/Odour				
Hardness				
Pump Drive				
Other Remarks				

Notes:- 1) Major use-Irrigation/domestic water supply (WS)/livestock watering, etc
      2) Pump Drive- Diesel/electric motor/gravity, etc
      3) Filling/Drying-state how often e.g. yearly, etc

Table 2: Other water sources (Boreholes, wells, shafts, well points, etc)

Type				
Grid Ref.				
Owner				
Permit No.				
Date Constructed				
Depth (m)				
Yield (m^3)				
Drying				
Major Use				
Ha (of Irr.)				
M^3 (of WS)				
Colour				
Taste/Odour				
Hardness				
Pump Drive				
Other remarks				

Notes:- 1) Major use- Irrigation/domestic water supply (WS)/ livestock watering etc
      2) Pump Drive- Diesel/electric motor/gravity, etc
      3) Filling/Drying-state how often e.g. yearly, etc

If you have more than four dams or sources, please use two or more sheets.

Integraal waterbeheer, instituties en welzijn onder druk: perspectieven van onderop uit Zimbabwe

Collin C. Mabiza

Samenvatting

De semi-aride omstandigheden in zuidelijk Afrika heeft tot gevolg dat de meerderheid van de bevolking beperkte toegang tot water heeft. Grote afhankelijkheid van regenafhankelijke de landbouw in delen van de regio, zoals het Limpopo stroomgebied, heeft geleid tot grootschalige voedselonzekerheid en armoede. Verbeterd waterbeheer is van essentieel belang daar het kan bijdragen aan het verhogen van de levenstandaard, het welzijn en sociaaleconomische ontwikkeling. In een poging om het waterbeheer te verbeteren hebben de meeste van de landen in dit stroomgebied het concept integraal waterbeheer (in het Engels: *Integrated Water Resources Management*, IWRM) geïntroduceerd. Zimbabwe was een van de landen die IWRM al in het vorige millennium hebben geïntroduceerd. Gezien de tijd die verstreken is sinds invoering van IWRM is het belangrijk om na te gaan of de waterbeheerpraktijken zijn verbeterd en hoe dit heeft bijgedragen aan een groter welzijn van mensen die leven in het stroomgebied. Deze analyse is essentieel voor het verbeteren van het IWRM concept en de ontwikkeling van nieuwe raamwerken voor waterbeheer.

Een benadering van onderop is gebruikt voor de analyse van lokale waterbeheerpraktijken in het Mzingwane deelstroomgebied dat onderdeel vormt van het Limpopo stroomgebied en dat gelegen is in het zuiden van Zimbabwe. Dit is een deelstroomgebied dat onder druk staat, zowel wat betreft de (natuurlijke) agro-ecologische, als de sociaal-politieke en economische omstandigheden. Het doel van deze studie was om te begrijpen wat de waterbeheerpraktijken zijn op lokaal niveau, en welke factoren deze praktijken vormgeven. Het theoretisch kader dat gebruikt is voor de analyse combineert de theorie van sociale constructie met het concept water controle en met een focus op het levensonderhoud van mensen (in het Engels: *livelihood*). Omdat waterbeheer het levensonderhoud van mensen zou verbeteren is voor een benadering van onderop gekozen. Deze studie analyseert verscheidene waterbeheerpraktijken om een breder begrip van de verschillende aspecten van het levensonderhoud van mensen te begrijpen. In de literatuur is er een trend om het welzijn van mensen gefragmenteerd te analyseren, bijvoorbeeld door alleen naar drinkwater te kijken, of naar het productief gebruik van water. Dit geeft een incompleet beeld van de lokale waterbeheerpraktijken. Deze studie is gebaseerd op een gevalsstudie benadering. De vijf gevalstudies hebben betrekking op de toegankelijkheid van water voor huishoudelijk en productief gebruik, inspanningen gericht op het behoud van levensonderhoud en milieuwaarden, waterbeheer voor de landbouw, geschillen over de stedelijke watervoorziening en stroomgebiedplanning. In alle gevallen werd gezocht naar een goed begrip van wat de waterbeheerpraktijken werkelijk vormgeeft.

De studie heeft eerst het integraal waterbeheer concept geanalyseerd. Deze kritische analyse werd ondernomen vanuit een conceptueel perspectief en vanuit een praktisch oogpunt met betrekking tot de invoering van het raamwerk. Conceptueel lijkt het integraal waterbeheer concept zich niet rechtstreeks te richten op het verbeteren van het levensonderhoud van mensen. Dit zou kunnen verklaren dat het ontwikkelen van water infrastructuur, dat het welzijn van mensen zou kunnen verbeteren, niet vaak onderdeel is van integraal waterbeheer plannen. Verder is het concept integraal waterbeheer niet helder over wat er geïntegreerd moet worden. Twee kritieken worden in dit verband vaak geuit. Enerzijds wordt verweten dat het concept onrealistisch is over wat er allemaal geïntegreerd kan worden- het wil teveel aspecten onder the noemer waterbeheer brengen. Anderzijds wordt verweten dat het concept niet integraal genoeg zou zijn en dat belangrijke aspecten vergeten worden. De problemen met het invoeren van IWRM hebben geleid tot een roep om een lichte versie van IWRM (*Light IWRM*).

De eerste gevalstudie probeert de volgende vraag te beantwoorden; wat zijn de waterbeheerpraktijken op lokaal niveau? The praktijken van watergebruikers van verschillende soorten en type van waterbronnen werden geanalyseerd, namelijk een put met een handpomp, een wetland en een waterput met een windmolen. De theoretische concepten "praktijk", "interactie" en *"bricolage"* zijn gebruikt om het locale waterbeheer te bestuderen, en voor de onderzoeksmethode werden focusgroep discussies, interviews en participerende observatie gebruikt. De bevindingen laten zien dat ondanks het feit dat integraal waterbeheer al ongeveer tien jaar geleden is geïntroduceerd in Zimbabwe, de waterbeheerpraktijken op lokaal niveau nog steeds plaatsvinden buiten dit raamwerk. De besturen van het stroomgebied en van de deelstroomgebieden zijn afwezig op het lokale niveauraad en beïnvloeden de lokale waterbeheerpraktijken niet. De studie vond dat waterbeheerpraktijken voornamelijk beïnvloed worden door de semi-aride omstandigheden en de sociaaleconomische context. Het droge klimaat beïnvloedt het gebruik van sommige water infrastructuur, en gecombineerd met een falende techniek, versterkt dit het besef van waterschaarste onder de gebruikers. Dit kon tot op zekere hoogte verklaren waarom in sommige gevallen regels die door lokale comités waren ingesteld werden gebroken. Een belangrijke bevinding van de studie was dat bij watervoorzieningen waar water ook werd gebruikt voor productieve doeleinden de infrastructuur doorgaans beter onderhouden werd dan bij voorzieningen die slechts gebruikt werden voor huishoudelijke doeleinden. De bevindingen van dit hoofdstuk zet vraagstekens bij de benadering die geheel gericht is op de vorming van nieuwe instituties voor integraal waterbeheer. Een alternatieve benadering om waterbeheer te verbeteren zou een focus op het verbeteren van het levensonderhoud van mensen kunnen zijn, bijvoorbeeld door het toegankelijk maken van water voor productief gebruik.

De tweede gevalstudie onderzocht hoe lokale actoren in hun levensonderhoud voorzien en wat daarin de rol van het milieu is. De context van deze studie werd gekenmerkt door negatieve sociaaleconomische en natuurlijke omstandigheden. Het onderzoek richtte zich op twee activiteiten die beiden een effect hebben op water, namelijk informele goudwinning en het beteugelen van geulvorming. De moeilijke sociaaleconomische omstandigheden en het droge klimaat dwongen lokale actoren om de natuurlijke hulpbronnen en institutionele kansen waartoe zij zich toegang konden verschaffen te gebruiken (i.c. misbruiken). Actoren verrichten schijnbaar tegenstrijdige activiteiten: ze participeren in milieuvriendelijke projecten terwijl ze tegelijkertijd het milieu aantasten, in één en hetzelfde microstroomgebied. Dit kan worden verklaard doordat non-gouvernementele organisaties (NGOs) projecten introduceerden die het doel hadden om het milieu te rehabiliteren door middel van Voedsel-Voor-Werk projecten (in het Engels: *Food-For-Work*, FFW), waardoor lokale actoren toegang kregen tot voedseldonaties. Maar toen een NGO stopte met de voedseldonaties kwamen de lokale mensen niet meer opdagen om de erosiegeulen te rehabiliteren. De studie concludeerde dat ondanks dat FFW projecten in het levensonderhoud van mensen proberen te voorzien, dit korte termijn acties zijn die de problemen waarmee de meerderheid van de kwetsbare wereldbevolking geconfronteerd wordt niet oplossen.

Het waterbeheer van de regenafhankelijke landbouw op veldniveau was de derde gevalstudie. De ervaringen van de verschillende actoren werden opgetekend en gebruikt om te analyseren hoe waterbesparende technieken worden gepromoot onder kleine boeren. Het pakket waterbesparende technieken waar het om gaat bestaat uit minimale grondbewerking (Engels: *minimum tillage*), beheer van de bodemvruchtbaarheid, en het bedekt laten van de bodem met organisch materiaal (Engels: *mulching*). Het uitvoeren van dit pakket was teveel gevraagd voor veel kleine boeren, die vooral klaagden over de grote hoeveelheid arbeid die nodig was. Toch probeerden de boeren de technieken uit te voeren. Ze deden dit voornamelijk om in aanmerking te komen voor gratis zaaigoed en andere inputs die NGOs beschikbaar stelden. Het hoofdstuk concludeerde dat kleine boeren zich inlaten met NGOs die waterbesparende technieken promoten omdat dit contact nuttig kan zijn voor hun levensonderhoud, en niet noodzakelijkerwijs omdat zij ervan overtuigd zijn dat duurzame landbouw (Engels: *conservation agriculture*) de gewasopbrengst kan vergroten.

De vierde gevalstudie analyseerde de manier waarop de politieke context de strijd over de controle van de stedelijke watervoorziening kan verklaren. Deze casus gaat over de landelijke overheid die van zins was de watervoorziening van de stad Bulawayo over te nemen. Dit resulteerde in een strijd tussen the gemeentelijke overheid en de bewoners enerzijds, en de landelijke overheid anderzijds. Bevindingen laten zien dat de centrale overheid de overname van de watervoorziening rechtvaardigde op grond van het verbeteren van de efficiëntie. Echter, centraal in deze strijd stond de symbolische betekenis van water. Voor de bewoners van de stad symboliseerde de watervoorziening van de stad trotsering: trotsering van de natuur vanwege het droge klimaat en trotsering van de landelijke politiek omdat de stad zelf de dammen voor de watervoorziening had aangelegd. Voor de landelijke overheid betekende de controle een grotere controle over de gemeentelijke autoriteiten. In tegenspraak met de adviezen van internationale financiële instellingen, kan economische efficiëntie op zichzelf niet verklaren waarom bepaalde waterbeheermodellen geaccepteerd worden en andere niet. In sommige gevallen worden waterbeheermodellen geaccepteerd of geweigerd op basis van de maatschappelijke waarde van water.

Het laatste empirische hoofdstuk analyseert processen in stroomgebiedplanning, en in welke mate de resulterende plannen aansluiten bij de welzijnsverwachtingen van watergebruikers. Daartoe werd de formulering van het ontwerpplan van het Mzingwane stroomgebied onderzocht. Het bleek dat belanghebbenden nauwelijks participeerden in het planningproces; zelfs bij het verzamelen van relevante gegevens speelden ze geen rol. Deelname werd gehinderd door de slechte sociaal-economische situatie toen het ontwerpplan werd gemaakt, ondanks dat de wetgeving en de beleidsrichtlijnen participatie van belanghebbenden voorschrijven. Beperkte deelname van belanghebbenden verklaart waarom het ontwerpplan niet de echte levensonderhoudvraagstukken heeft meegenomen, zoals toegang tot water voor zowel huishoudelijke als productieve doeleinden.

De algemene conclusie van deze studie is dat ook al veronderstelt de invoering van IWRM dat waterbeheer kan worden verbeterd door het overnemen van internationaal geaccepteerde raamwerken, zijn het vooral lokale sociaal-politiek-economische en fysieke factoren die water beheer vormgeven. Daarom kan het waterbeheer niet verbeterd worden in isolatie van het verberen van het welzijn van watergebruikers.

Curriculum Vitae

Collin Mabiza was born on 6 November 1972 in Chivhu, Zimbabwe. In 2001 he obtained a Bachelors degree (Honours) in Geography and Environmental Sciences from the University of Zimbabwe. The following year he enrolled for a Master's degree in Environmental Policy and Planning at the same university. After graduating in 2004 he joined Chinhoyi University of Technology's Department of Environmental Science as a lecturer. In November 2005 he was accepted for a PhD research position in the Challenge Programme on Water and Food (CP 17). The University of Zimbabwe appointed him as a Research Associate in the Department of Civil Engineering, a position which facilitated his research. As a member of the Department he taught several courses of the regional WaterNet Master programme in IWRM, and coordinated the modules on Socio-Economics of Water and Environmental Resources and on Conflict Resolution. His career to date has focused on the livelihoods-resource management nexus. Collin is happily married to Nana.

T - #0410 - 101024 - C20 - 240/170/11 - PB - 9781138000360 - Gloss Lamination